JOURNAL OF SEMITIC STUDIES
MONOGRAPH NO. 1

NEW LIGHT ON THE LIFE OF MUHAMMAD

BY
ALFRED GUILLAUME
Emeritus Professor of Arabic in the University of London

MANCHESTER UNIVERSITY PRESS

Printed in Great Britain at the University Press, Cambridge
(Brooke Crutchley, University Printer)

CONTENTS

NEW LIGHT ON THE LIFE OF MUHAMMAD

INTRODUCTION

Attention was first called to the existence of a manuscript containing a report of Ibn Isḥāq's lectures on the Life of Muhammad by Dr Muḥammad al-Hajawī in 1932.[1] A few years later mention was made of it by the late Dr Aḥmad Amīn in his *Ḍuḥā'l-Islām*.[2] The former had seen the MS. itself; the latter had merely heard of it.

It is important to note that the work is not a record of Ibn Isḥāq's lectures alone. The title-page of the beginning of the work has not survived, but Part II is entitled "Part II of the 'Book of the Campaigns' (*Maghāzī*). The report of Yūnus ibn Bukayr from Muḥammad ibn Isḥāq and others. The report of the shaykh Abū'l-Ḥusayn Aḥmad ibn Muḥammad ibn al-Naqūr al-Bazzāz from Abū Ṭāhir [Muḥammad ibn 'Abdu'l-Raḥmān][3] al-Mukhliṣ from Riḍwān from Aḥmad ibn 'Abdu'l-Jabbār al-'Utāridī from Yūnus. May God be pleased with them all."

Thus, though all the traditions which the MS. records come from one who heard Ibn Isḥāq's lectures at Kūfa, the same reporter drew more than two hundred traditions from other sources. It will be seen that there are only four intermediaries between Yūnus and al-Bazzāz. Yūnus died in 199 A.H., almost fifty years after the death of Ibn Isḥāq. Al-'Utāridī was born in 177 and died in Kūfa in 271 or 272.[4] I have not succeeded in finding any reference in the Arab biographies to the other intermediaries, but if we could allow a fair lapse of time between each of the other intermediaries we should arrive at a date round about, or after, the year 400 for al-Bazzāz. The first date mentioned in the MS. is 456, so that we have a well-authenticated document scrupulously handed on from one who attended the lectures of the earliest extant authority on the life of the prophet.

It had been my intention to publish the text of the MS. with a translation and bibliographical notes; but I have reluctantly come

[1] *Majalla* of the Arab Academy of Damascus (1932), XII, p. 110.
[2] Cairo, 2 vols. (1351–3/1933–6). Cf. Brockelmann, *G.A.L.*, S.I., 206.
[3] The words in square brackets are taken from the *isnād* on fo. 19 and fo. 41.
[4] *Tahdhīb al-Tahdhīb* (Haydarabad, 1325/7), I, pp. 51–2.

to the conclusion that I must leave this exacting task to a younger man. Occasionally words or phrases in the photostat are illegible, and a visit to the Library of the Qarawīyūn mosque at Fez may well be necessary to establish, or to endeavour to establish, what a particular reading actually is. But fortunately nothing of real importance is illegible. Omissions which would be regrettable if not culpable in an edition of an Arabic text, which must faithfully reproduce what the writer said, can legitimately be passed over by one who seeks to put into the hands of scholars an account of the contents of this valuable work. Therefore I have contented myself by trying to do for Yūnus b. Bukayr what Wellhausen did for al-Wāqidī in his *Muhammad in Medina*. This translation in German has been used by all subsequent writers on the life of the prophet, and will not be supplanted until the Arabic text has been printed.[1] With the traditions which Ibn Bukayr culled from other traditionists I am not concerned; but here again the task of tracking down these additional traditions in the vast literature of Islamic *ḥadīth* is daunting; not until it has been undertaken can it be determined whether the MS. adds to our knowledge of traditions current in the second half of the second century of Islam. It would be surprising if it did not contain anything new from some of the authorities quoted by Ibn Bukayr. At the least they would be important inasmuch as they would give an earlier date to some of those traditions that are extant in other works.

In spite of the fact that a fairly full summary of the contents of the MS. follows below it may be worth while to give a brief résumé of the principal additions which this MS. makes to our knowledge of the life of Muhammad and of Arabian society in his time. Specialists in the antiquities of South Arabia may find a little to interest them in the material. Interesting in the anthropological field are the stories which are told to illustrate the potency of curses uttered in rhyme (*saj'*). Ibn Isḥāq introduces them with a comment to the effect that truly effective curses should be pronounced in *saj'*.[2]

Ibn Isḥāq's comments are valuable in that they transmit to us the ideas and beliefs that were current in the Ḥijāz in the early years of the second century of the Hijra. For instance, his view of the influence of monotheism was that it was merely superficial;

[1] Dr J. M. B. Jones tells me that he has every hope that his edition of the text will appear this year.

[2] Goldziher gave a masterly exposition of the subject in his *Abhandlungen zur arabischen Philologie* (Leyden, 1896).

the Arabs were illiterate and what they heard from Jews and Christians had no effect on their lives. It must be remembered that he was talking about Western Arabia, and one would have thought that the influence of the synagogue or synagogues in Medina and its suburbs would have been considerable, especially when one bears in mind the close agreement between the Koran and the Talmud in teaching and terminology. It is not unlikely that Ibn Isḥāq inherited a poor view of Christian influence in the Ḥijāz from his grandfather, who was a Christian from ʻAyn al-Tamr and was brought as a slave to Medina by Khālid b. al-Walīd.

One of the few stories about Muhammad's life before his call is the account of his having placed the black stone in its place when the Kaʻba was rebuilt. The MS. records a tradition purporting to come from the prophet's poet laureate Ḥassān b. Thābit to the effect that he himself saw ʻAbdu'l-Muṭṭalib put the stone in its place.

Another well-known story about the prophet's youth is given us in what must have been its original form. A certain Zayd b. ʻAmr upbraided him for eating meat that had been sacrificed to idols and condemned idols as futile and impotent. This admonition had such an effect on Muhammad that he said that never again did he offer sacrifices to idols nor show them any honour. Unfortunately the MS. does not tell us what Zayd's religion was. It simply says that he followed the religion of Abraham but was neither Jew nor Christian. This of course may be true. We know how the horrible persecutions and atrocities which were perpetrated by various sects of Christianity on one another finally drove out all but a minority of Arabs from the Christian fold, and it is possible that similar deplorable conflicts between Jews and Christians, which in the Yaman culminated in wars and massacres, may have made men who accepted the fundamental rightness of the two religions stand aloof from both. We know very little about Christianity in the Ḥijāz at that time. It may be that the references to some members of the tribe of Quraysh, who in the pagan period withdrew from the haunts of men and fed the poor who came to them, indicate an imitation of Christian monasticism, because there is nothing in the pagan religion of Arabia to encourage the belief that this was an indigenous development, nor can it owe anything to Judaism.

Another tradition which illustrates the beliefs of the time is that Muhammad suffered from the evil eye and his wife Khadīja used to send for an old woman in Mecca to charm it away. When the Koran came down Muhammad dispensed with her services.

The story of Muhammad's temporary concession to Meccan polytheism is given, but as we know from Ṭabarī that Ibn Isḥāq recorded it, nothing material is added to our knowledge.

To the best of my belief the MS. contains the only record of a Jew holding and cultivating land in Mecca. As one would expect, the land contained palm-trees and a well.

A long section dealing with the fate of a man who attempted to murder his Arab companion and showing how his illicit visits to the wife of the Abyssinian king brought about his punishment and subsequent death is not entirely new, though in its completeness it is found only here. The inference is that a fairly large number of Arabs were familiar with the Abyssinian language, and that intercourse between Arabs and Abyssinians must have been considerable in those days.

Lastly it may be noted that the MS. carries on a few incidents into the period of the caliphate.

A comparison of the text of the MS. with the edition of Ibn Hishām forces me to admit that in one respect at least I failed to do justice to Ibn Hishām's work as an editor.[1] If Ibn Isḥāq gave his lectures in the form and order in which Yūnus b. Bukayr recorded them—naturally we have no information on the point—then we owe much to Ibn Hishām for his painstaking efforts to introduce some sort of logical and chronological order into the narrative. As the full summary will indicate, the pages of the MS. run from subject to subject at times, and without the *textus receptus* to keep one on the right track it would be a difficult task to arrange the material in a way satisfactory to any reader. Men like Ibn Hishām and more especially Ṭabarī have performed a service of inestimable value in introducing order into what may have been an incoherent assembly of traditions. However, as has been said, we do not know what form Ibn Isḥāq's lectures took, for the simple reason that his work has not survived. And in this connexion it is important to remember that the MS. is not confined to a transcript of Ibn Isḥāq's lectures but contains a great deal of extraneous material. From this it follows that it cannot be a straightforward report of the original authority's words, though one would expect the biography to follow his order had

[1] A. Guillaume, *The Life of Muhammad*, with Introduction and Notes (Oxford University Press, 1955), p. xli.

he arranged his lectures on a logical method. On the whole it seems safer to leave the question open.

At any rate this manuscript is of such great importance and interest that one hopes that in the not too distant future some younger scholar will be able and willing to publish its text. For the Arabist it has much of interest to offer in variant readings, archaic words, and names different from those given by Ibn Hishām.

THE MANUSCRIPT

The MS. in the Qarawīyūn library at Fez is numbered 727. No good purpose would be served by my anticipating the labours of a future editor of the text, but a little ought to be said about the MS. itself. By great good fortune a former colleague of mine at the School of Oriental and African Studies, Dr J. F. P. Hopkins, while pursuing his researches in Morocco, was able to look at the MS. and he very kindly gave me some notes about it, and a transcript of several of its pages. He wrote: "The MS. as it exists consists of a single *kurrāsa* of about twelve folios, and a complete volume of about four equal parts, each of about eighteen folios. It is written on paper about 8 in. by 6 in. The whole is very much worm-eaten, but it is quite usable and on the whole very legible, especially the apparently older hands.

The folios are numbered, but the writing is so faded as to be practically illegible, the worms having eaten their way through the place where most of the numbers were. Moreover, they do not seem to be in order, though as far as I can see the text itself follows on correctly. The work is so short and the chapter-headings so frequent, that this is of little consequence. The text is not all in the same hand, and some appears to be later than the rest, though all seems to be ancient. The points on letters are often missing. Some hands write *qāf* and *kāf* in the Maghribī style, others not."

The MS. has been most carefully gone over by the transmittors, and the validating word *ṣaḥḥa* is fairly frequently used. This work must have been highly prized in ancient days before it was neglected and left to waste away, for it has copious *ijāzas*. Folio 33 states that the work consisted of seventeen parts. Of these only three have survived intact with rather more than a half of the first part and with a fragment of Part v. The Summary, with its references to the parallel version of Ibn Hishām, will show how very difficult it is to compare the size of the MS. with the latter.

The number of its pages does not help because the MS. contains much that has nothing to do with Ibn Isḥāq and, as has already been pointed out, the order of the narrative is not that of Ibn Hishām. Were I to venture a rough guess I should be inclined to say that the MS. contains about a fifth to a quarter of the *Sīra* as we have it in Ibn Hishām's version.

It is a pleasure to record my thanks to Dr I. S. Allouche for kindly obtaining a microfilm of the MS. for my use.

SUMMARY OF THE CONTENTS

No good purpose would be served by a complete translation of the manuscript, because much of it runs parallel with the *textus receptus* as we have it in the version of Ibn Hishām which is based on Ziyād ibn 'Abdullāh al-Bakkā'ī's report of Ibn Isḥāq's lectures. Where there are minor differences between the two texts they concern the Arabist rather than the English reader. The importance of this manuscript lies in those passages which restore to us material that Ibn Hishām omitted from his text for the reasons which he has given in his Introduction to his edition, and with this material the pages that follow are chiefly concerned.

To facilitate a comparison between the two reporters of Ibn Isḥāq's lectures, that is to say Yūnus b. Bukayr and al-Bakkā'ī, references to the *editio princeps* published by Wüstenfeld a century ago[1] are given wherever the two authorities overlap, and a reference to my translation of the latter[2] is added for the benefit of those who are not familiar with Arabic. Other authorities who have used Ibn Bukayr's work and who are cited are: al-Ṭabarī, *Annales quos scripsit Abu Djafar Mohammed ibn Djarir at-Tabari* cum aliis edidit M. J. De Goeje (Leyden, 1879–90); Abu'l-Fidā' Ismā'īl b. 'Umar b. Kathīr, *Al-Bidāya wal-Nihāya* (Cairo, 1348/1929); and Abū'l-Qāsim 'Abdu'l-Raḥmān ibn 'Abdullāh al-Suhaylī, *Al-Rauḍ al-Unuf* (Cairo, 1914).

Ibn Hishām is cited as I.H., my translation as L., al-Ṭabarī as T, Ibn Kathīr as I.K. and al-Suhaylī as S.

[1] *Das Leben Muhammed's nach Muhammed Ibn Isḥāq*, bearbeitet von Abdelmelik Ibn Hischām (Göttingen, 1856–60).

[2] *The Life of Muhammad* (Oxford University Press, 1955).

[*Part I*]

The Story of Tubba' the Ḥimyarite

MS. fos. 1b–3a; I.H. pp. 12–18; L. pp. 6–11

Here we are given the antiquarian note that Tubba' camped in Wādī Qubā where he dug a well which is known to this day as "The King's Well".[1] An Ausite named Uḥayḥa ibn al-Julāḥ ibn Jaḥjaba ibn Kalada[2] and a Jew named Benjamin of Qurayẓa went out from Medina to negotiate with Tubba' when he was about to attack the city. [This Uḥayḥa was the first husband of Salmā, a woman of the tribe of al-Najjār. Later she married Hāshim and so was the prophet's great-grandmother.] When Uḥayḥa came to the king he said "We are thy people". The significance of this, as I have shown elsewhere,[3] is that it was Ibn Isḥāq's set purpose to show that the prophet belonged to Medina by descent through his great-grandmother, as well as to Mecca, and that it was the Anṣār who supported him when his fellow-townsmen persecuted him. Furthermore, the Anṣār were of one blood with the kings of the south. In I.H. p. 13 (L. p. 7), it is said that the Anṣār used to fight Tubba' by day and provide him with food by night at which he exclaimed "Truly our people are generous".[4] Ibn Bukayr reports that this Benjamin (and not the two rabbis of Ibn Hishām) frightened the king away from Medina by telling him that it would be the home of a prophet who was to come from Quraysh. Tubba' left at once because the report of a devastating fire in the Yaman reached him. The "poem" which he composed must rank as one of the worst in all Arabic literature. One would be tempted to think that the author had written thus because he was putting Arabic into the mouth of a Yamanite, were it not that the bulk of the poetry recorded in the *Sīra* is puerile and obviously forged by someone living long after the events with which it is ostensibly related.

[1] So Muḥammad b. Maḥmūd b. al-Najjār, *Akhbār Madīnat al-Rasūl*, ed. Ṣāliḥ Muḥammad Jamāl (Mecca, 1947), p. 37; and *Yācūt's Geographisches Wörterbuch*, ed. F. Wüstenfeld (Leipzig, 1866), I, 432 (*sub* Bi'r Rūmah). Both Ibn al-Najjār and Yāqūt refer to the *ahl-al-siyar* as authorities for this statement.

[2] I.H. p. 88, L. p. 59 "Kulfa". [3] *The Islamic Quarterly*, I (1954), 1–11.

[4] One of my reviewers would translate *qaumanā* by "our enemies"; but the whole tenor of Ibn Isḥāq's writing leaves no doubt of his insistence on the southern kingly origin of the two tribes of the Anṣār. By calling the men of Medina "our people" the Tubba' is acknowledging their claim.

Both recensions contain the story that Tubbaʿ intended to sack Mecca and rob it of its treasures. Ibn Bukayr adds the detail that God sent a scorching wind which caused his whole body to shrink, and healed him when he abandoned his design. It is said that the place where the Hudhaylīs met Tubbaʿ was al-Duff, between Amaj and ʿUsfān.[1] Both sources agree that Tubbaʿ behaved more or less as a Muslim pilgrim would, but Ibn Bukayr adds that he intended to take the black stone away with him to the Yaman; however he gave up the project when he saw the determined opposition of Quraysh headed by Khuwaylid who was later to be the prophet's father-in-law.

After stating that Tubbaʿ arrived in the Yaman with his forces Ibn Bukayr reports that the Yamanīs had two towns, Ma'rib and Ẓafār. The king's palace in Ma'rib was plated with gold; his palace in Ẓafār was built of marble. He spent the summer in Ẓafār and the winter in Ma'rib. The royal princes were educated in Ma'rib where they learned oratory (*manṭiq*).[2] A pillar from the holy city[3] was inscribed at the top in the ancient writing with the words "To whom belongs the kingdom of Ẓafār?" etc.[4]

The story of the fate of the devil who inhabited the Yamanī temple is very different and would be unintelligible without Ibn Hishām's version. Ibn Bukayr reports that the Yamanīs had built a golden temple for a *shayṭān* whom they worshipped. They used to set before him a menstruous garment and cut the throat of an animal (*dhabaḥa*) for it so that the blood ran into the garment. He would come out and receive the blood and speak to them and they would ask an oracle of him. Yāqūt[5] quotes this tradition with the variant "they used to soak a menstruous garment with the blood of the sacrifice for him".[6]

The Jews, after getting the king's promise that if they made the *shayṭān* come out he would accept the Jewish faith, sat down with their scriptures and recited the names of God. Thereupon the *shayṭān* came out openly and fell into the sea in the sight of all.

[1] Cf. Al-Azraqī, *Akhbār Makka* (Mecca, 1352), I, 79f. for yet another version of Ibn Isḥāq's lectures.

[2] It is unlikely that the writer meant "logic" in the later sense of the word.

[3] *balad al-ḥarām*.

[4] Cf. I.H. p. 47, L. p. 34 for some variants.

[5] *Op. cit.* III, 882.

[6] See further W. Robertson Smith, *The Religion of the Semites*³ (London, 1927), pp. 447f. This practice is not recorded there.

The Killing of Tubba‘ and the Abyssinian invasion of the Yaman

MS. fos. 3a–4b; I.H. pp. 18–29; L. pp. 10–21

This passage has suffered severe compression. The reader would suppose that the Tubba‘ who was murdered was the one spoken of above, but Ibn Hishām makes it clear that it was his son. Here it is said that the princes were enraged because the Tubba‘ had kept them away from their families on a long campaign and had attacked their religion and insulted their ancestors. As they were about to kill the Tubba‘ he asked that he might be buried upright because sovereignty would not depart from them while he stood on his feet. But when they killed him they swore that he should not rule them alive or dead, and so they buried him head downwards. Dhū Hamdān refused to take part in the assassination. Ibn Hishām calls this man Dhū Ru‘ayn, and he is so named in the two couplets quoted in both recensions. The name of the Tubba‘'s brother is given as ‘Abdu Kulāl as against Ibn Hishām's ‘Amr. A lament on his unfortunate brother which he is said to have composed follows.[1] In this version it is the Jews, not the physicians and diviners, who advise the new king to execute the nobles who incited him to fratricide.

This section jumps to the visit of Daus to the emperor. The details of the story of what happened at the court are more vivid, and incline one to regard this version as superior in some respects to that of Ibn Hishām. It is asserted that the Abyssinian king was subject to the emperor. Here the name of the commander of the Abyssinian invaders is Rūzbah as against Ibn Hishām's Aryāṭ. The attacks of the Yamanī cavalry were so dangerous that the invaders were in danger of annihilation. Rūzbah accused Daus of bringing him there in order to get him killed and threatened to kill him first. Daus pointed out that the Ḥimyarites were formidable only when mounted, and advised the Abyssinians to throw their shields before them so that the horses would stumble and throw their riders. These tactics were adopted, the Ḥimyarites were defeated and the Abyssinians occupied Ṣan‘ā. The two recensions agree in describing the treachery by which Abraha slew his rival and became viceroy of the Yaman.

[1] It has no relation to the anonymous elegy given in I.H. p. 18, L. p. 12.

The Story of the Elephant

MS. fos. 4b–6a; I.H. pp. 29–41; L. pp. 21–30

The broad outline is the same, though Ibn Bukayr's account is more lively and adds some details to Ibn Hishām's account, e.g. that the cathedral in Ṣan'ā had golden domes. He reports that the southern tribes of 'Akk and Khath'am with the Ash'arites marched out with Abraha chanting as they went:

> The town is a town to be destroyed (lit. "eaten")
> 'Akk and the Ash'arites and the Elephant will destroy it.

Ibn Hishām says that Khath'am and its allies fought against Abraha, and joined him only after he had defeated them. Nufayl, their leader, was captured and escaped death by volunteering to act as guide to the Abyssinians. Ibn Bukayr reports that while on his march Abraha sent one of the Banū Sulaym to summon men to pilgrimage to Ṣan'ā, and that one of the Ḥums of Banū Kināna met him and slew him, and that this added to Abraha's fury and haste. Further, that it was the men of al-Ṭā'if who supplied a guide from Hudhayl by the name of Nufayl.[1] Ibn Hishām also says that al-Ṭā'if supplied a guide named Abū Righāl, and that he died in al-Mughammas.

According to Ibn Hishām, 'Abdu'l-Muṭṭalib interviewed Abraha through the good offices of Dhū Nafr, an acquaintance of his who was held prisoner in the Abyssinian camp. The latter was a friend of the keeper of the elephant. In the MS. the interview took place through an anonymous Ash'arite chamberlain of Abraha. The night before the attack on Mecca the very stars appeared to warn the invaders of approaching punishment. Their guide left them and entered the sanctuary. The Ash'arites and Khath'am broke their lances and swords and became guiltless before God of assisting in the attempted destruction of the Ka'ba. The birds which brought the stones were said to be black like *naḥāmīm* (flamingoes).

This is one of the stories which have been touched up by the narrator. In place of Ibn Hishām's version which says that Nufayl took hold of the elephant's ear and warned him that Mecca was a town inviolate, the Abyssinians are said to have sworn to the elephant that they would not lead him against Mecca, and after their repeated oaths the huge beast flapped his

[1] Evidently traditions are at variance here.

great ears in acceptance of their promise and loped off eagerly in the direction of home only to be turned back again.[1]

The Opening of the Well Zamzam and its Disputed Ownership

MS. fos. 7a–8a; I.H. pp. 92–4; L. pp. 63–4; I.K. II, pp. 245–6

The beginning of this story is missing and the fragment begins with the words "What do you think?" At first there is no material difference between the two *riwāyas*, though the wording is not always identical. Whether Zamzam, Ishmael's well, was produced by God or whether it was hollowed out by Gabriel's heel is disputed. Possession of the well is said to have added greatly to 'Abdu'l-Muṭṭalib's prestige and dignity and men came to the well to get a blessing from it. Ibn Bukayr's account is much shorter than Ibn Hishām's. There is no mention of visions, though divine guidance in finding the well is recorded. When lots were cast to see who should have the golden gazelles and swords that were found in Zamzam, 'Abdu'l-Muṭṭalib gave vent to the lines of rough verse cited by Ibn Kathīr.

The Potency of Invocations pronounced in "saj'"

MS. fos. 8a–9b; not in I.H.

I have been unable to find this passage elsewhere. Ibn Isḥāq says that Quraysh and the other Arabs in the pagan period when they desired to be vehement in imprecations uttered them in rhymed prose[2] and language that found its mark.[3] It is alleged that such imprecations were seldom fruitless. Some anecdotes are collected here to illustrate their power. The first rests on the

[1] At this point the narrative is out of order. The first of the interpolations from traditionists other than Ibn Isḥāq is found on fo. 6a which continues on fo. 6b with the note on the death of the prophet's mother (I.H. p. 107, L. p. 73), and the disinterment of 'Abdullāh ibn al-Thāmir (I.H. p. 25, L. p. 18). Ibn Bukayr's version is that a man dug up the grave in Ṣan'ā, whereas Ibn Hishām reports that it was in Najrān as one would expect. Ibn Isḥāq adds "he followed the religion of Jesus".

[2] See my *Prophecy and Divination*, London, 1938, pp. 245–50 *et passim*.

[3] *alqau l-kalām*. This expression does not appear to be noted in the Arabic lexicons. *Ilqā'* is used of satanic suggestion, and Sūra 54, 25 "Has the Reminder been cast upon him (*ulqiya 'alayhi*) alone among us" suggests that the verb has supernatural associations. Some such meaning as that given above or, perhaps, "made their words ominous" would seem to meet the case.

authority of a man who had lived in the pre-Islamic period, ʿAbdullāh b. Kharrīt by name. The traditionist states that every clan of Quraysh had its own meeting place (*nādi*) in the sacred mosque, and the Banū Bakr also had a place where they sat (*majlis*). Suddenly a young man burst in at the door and clung to the curtains of the Kaʿba. He was followed by an old man who went up to him to seize him, and as he did so his hands withered. When Quraysh learned from him that he was one of the Banū Bakr they told him in no uncertain terms that he was unwelcome. The young man told them that when his mother was left a widow she took her children to the temple and told them that if she died and they were wronged they were to go and take refuge there, for the temple would protect them. When the woman died the old man practically enslaved them for some years. On this occasion he had brought the youth with a herd of camels which he was going to sell, and when the young man saw the temple he remembered his mother's words and sought refuge there. The Quraysh tied the old man to his camel and sent him off with curses. [1]

The second story is concerned with a remarkable escape from a bandit. The tradition is reported from Abū Bakr who said that as he was travelling through a pass when he was a merchant, a man called out to him, and when he went up to him he found that he had been bitten by a snake and was unable to rise from the ground. The man asked him to bring him to his family who were at the foot of the pass, and so Abū Bakr mounted him on his camel and brought him to them. One of them exclaimed "You have indeed been helped by God in my opinion![2] There is no bandit more dangerous than this man." Abū Bakr went on to say that his camel led him astray and wandered off, and when he despaired of recovering it he lay down beside his saddle and wrapped himself in his garment and went to sleep. He was awakened by the feel of the camel's lip nuzzling his foot. He got up and rode off on it. (It is difficult to see the relevance of this story.)

The third story dates from the caliphate of ʿUmar. He was reviewing the men according to their register when an old man blind and lame went by being dragged violently by his guide. ʿUmar exclaimed that he had never seen a more unpleasant sight. One of the men present explained that the cripple was a son of Ṣabghā' cursed by Burayq, a nickname of a certain ʿIyāḍ. ʿUmar

[1] The language of the story is archaic.
[2] *maṣnūʿan laka*. For this idiom cf. T. I, 1924, 6.

ordered 'Iyāḍ to be summoned to his presence and asked him to tell him what had happened. 'Iyāḍ protested that it was something that had occurred in the pagan era, but 'Umar said that there was no harm in good Muslims talking about such matters. 'Iyāḍ then went on to relate that he had been left with the ten sons of Ṣabghā' who were kinsmen, and they squandered his property and kept him in subjection. When he appealed to them in the name of God and kinship and the right of protection they would not listen to him. He continued: "So I gave them a respite until the holy month came round. Then I lifted up my hands to Allah and said 'O God I implore Thee with an earnest prayer, slay the sons of al-Ṣabghā'[1] all but one, then smite his foot and leave him lame. When he is led about blind may trouble be his guide!'[2] Nine men died in one year one after the other and God smote the leg of this man and blinded his sight, and you see what his leader has to cope with." 'Umar said "This is indeed remarkable!"

One of those present claimed that the case of Abū Taqāṣuf the Khanā'ite of the tribe of al-Hudhayl was even more astonishing. They had treated a protégé in the same way as 'Iyāḍ had been treated, and he too endured until the holy month came round and then lifted up his hands and said: "O God, Lord of every believer and God-fearer, Who hearest the prayer of him who calls.[3] The Khanā'ite Abū Taqāṣuf has denied me my right and been unjust, so assemble for him the finest grains of sand between Qirān and al-Tawāṣuf.[4] They went down into an old well of theirs in the area he described when it collapsed upon them, and lo it is their grave to this day."

Another man claimed that the affair of the Banū Mu'ammal of Banū Naṣr was even more astonishing. The facts are the same as in the two preceding stories, except that one member of the Banū Mu'ammal named Riyāḥ had supported the man who had been wronged in his claim to justice. In the holy month he said: "O God, remove these men from the Banū Mu'ammal and bring dire punishment on their heads with a rock or the swords of a large army, saving only Riyāḥ who had no part in the wrong." As they were on their way as pilgrims they encamped at the foot of a mountain and God let loose a rock from the top of the mountain which carried stones and trees in its descent so that it crushed

[1] Previously Ṣabghā' has been written without the article.
[2] This and the curse that follows are in *saj'*.
[3] *kullu hātif*.
[4] These places are not mentioned by Yāqūt.

them at one blow, except Riyāḥ and his family who were in his tent, because he had done no wrong. 'Umar, much astonished, asked their opinion of these happenings, but the man said that he knew best, and he went on to explain that in the pagan era men knew nothing of a Lord, the sending of prophets, resurrection, paradise or hell, and God answered the prayers of the wronged against the oppressor to protect thereby one from another. And when God sent his apostle and they knew God and the sending of prophets and the resurrection and paradise and hell God said: "Nay but the Hour is their appointed time and the Hour is more disastrous and bitter",[1] so there were intervals of time and forbearance.[2]

'Abdu'l-Muṭṭalib's Vow

MS. fos. 9b–13a; I.H. pp. 97–100; L. pp. 66–8; and I.H. p. 69; L. p. 707; I.K. II, 248

Apart from small verbal variants there is little to call for notice in this section. The names of 'Abdu'l-Muṭṭalib's sons which are given by the editor Ibn Hishām on p. 69 are written here in a different order. It would appear from Ibn Hishām's list that 'Abbās was the eldest whereas our MS. puts al-Ḥārith first, 'Abbās seventh, and Ḥamza eighth. There is an interesting difference at one point: Ibn Bukayr's version states that it was a woman who was in charge of the divining arrows at Hubal's image, though it agrees with Ibn Hishām's version that a man presided on this occasion. There is an additional detail that the name of the woman soothsayer in Khaybar was Sajjāḥ (so S. I, 103). It is said that the Quraysh used to slaughter their sacrifices by the idols Isāf and Nā'ilah. And further that 'Abbās dragged 'Abdullāh from beneath his father's foot, and that in consequence his face was so deeply scratched (by his father's sandal) that he bore the mark to his dying day. The MS. confirms the statement of Ibn Hishām that the narrative was interspersed with *rajaz* verse which every authority on poetry regarded as spurious. Ibn Bukayr gives the text of ten poems uttered by 'Abdu'l-Muṭṭalib between each of the throws.

[1] Sūra 54, 46.
[2] *imlā'* may mean delay or postponement.

The Marriage of 'Abdullāh to Āmina and the Woman who made Overtures to Him

MS. fos. 13b–14b; I.H. pp. 100–1; L. pp. 68–9; I.K. pp. 49f.

Here there are verbal differences but the narrative is the same. The name of the woman who made overtures to 'Abdullāh is said to have been Umm Qibāl. She and he address each other in verse after 'Abdullāh's marriage to Āmina. Another poem attributed to 'Abdu'l-Muṭṭalib follows.

The Birth of Muhammad

MS. fos. 14b–16a; I.H. pp. 102–3; L. pp. 69–70; T. pp. 998–9; I.K. pp. 249–50; S. p. 106

The *saj'* of the supernatural announcement differs somewhat, and in place of the statement in Ibn Hishām that Āmina saw a light come forth from her by which she could see the castles of Buṣrā in Syria Ibn Bukayr has: "The sign of this shall be that a light shall come forth with him and fill the castles of Buṣrā in Syria, and when he is born call him Muhammad for his name in the Taurāt is Aḥmad whom the inhabitants of heaven and earth will praise. His name in the Gospel is Aḥmad whom the inhabitants of heaven and earth will praise, and his name in the Furqān is Muhammad, so call him by that name."

Alongside the statement that 'Abdullāh died while Āmina was pregnant Ibn Bukayr reports that there was another tradition that 'Abdullāh lived until Muhammad was twenty-eight months old. This tradition is to be found in S. and T.

The MS. agrees with T. in saying that 'Abdu'l-Muṭṭalib presented the child to Hubal in the middle of the Ka'ba, a detail which Ibn Hishām dropped. Then follow five pieces of *rajaz* attributed to 'Abdu'l-Muṭṭalib. One of them is given by S. and I.K.

Muhammad was born in the year of the Elephant, and was twenty years old in the year of 'Ukāẓ.

The Prophet's Foster-mother. Two angels Split his Belly and Extract Something Therefrom

MS. fos. 15b–17b; I.H. pp. 103–6; L. pp. 70–3; I.K. pp. 261, 275

The differences in the narration here are of philological rather than general interest.

The Death of 'Abdu'l-Muṭṭalib

MS. fo. 18a; I.H. pp. 108–10; L. pp. 73–6

Here the MS. shows that the order of the narrative has been disturbed. The story of the Tubba' which I have marked 1b follows on the line after the cleansing of the child Muhammad's heart. The "poems" ascribed to 'Abdu'l-Muṭṭalib's daughters differ in the order given, and they are much shorter in the MS.

Umayma has two lines as against Ibn Hishām's seven; 'Ātika four instead of eight; Ṣafīya five instead of twelve; al-Bayḍā' four instead of nine. Possibly a page is missing as two more daughters should have had their say. The next section is marked as Part II.

Part II

The Story of Baḥīrā

MS. fos. 19a–21b; I.H. pp. 115–17; L. pp. 79–81; I.K. pp. 283–99; S. pp. 119–20

Apart from small textual differences and occasional additions this story runs on the same lines as the *textus receptus*. The MS. contains three poems attributed to Abū Ṭālib; part of the second is to be found in S.[1] After the words "the apostle of God grew up, God protecting him and keeping him from the vileness of heathendom (MS. "and its vices") because He wished to honour him with apostleship" the MS. has "while he followed the religion of his people". T. (1126) records the tradition from 'Alī that twice the prophet was tempted to spend the night with women in Mecca (the wording in the MS. is slightly more blunt) and on each occasion Allah sent him to sleep so that he did not wake until sunrise. Ibn Kathīr's account is practically the same as the MS. He got it from al-Bayhaqī who stigmatizes the tradition as *gharīb jiddan*.

Khadīja daughter of Khuwaylid

MS. fos. 21b–22a; I.H. pp. 119–21; L. pp. 82–3; I.K. pp. 293–5

Here the MS. carries back Khadīja's genealogy on the female side two generations beyond the grandfathers named in Ibn Hishām, and her grandmothers to the eleventh generation as against Ibn Hishām's three generations. It adds the note that all

[1] The other two I do not find elsewhere.

Muhammad's children by Khadīja were born before revelation came to him and places them in the order: Zaynab, Umm Kulthūm, Ruqayya, Fāṭima, al-Qāsim, al-Ṭāhir, and al-Ṭayyib. The three sons died before Islam and the daughters survived and migrated with the prophet.

This section omits all mention of Khadīja's conversation with Waraqa and his assertion that Muhammad was the prophet of this people as well as the poem attributed to him in Ibn Hishām.

The Story of the Rabbis

MS. fos. 22b–23b; L. pp. 90–5; I.H. pp. 129–36

Here the MS. differs widely from Ibn Hishām. There is no question of variant readings: we are given a different "historical" introduction to the traditions which the author proceeds to cite. Most probably when Ibn Isḥāq lectured he allowed himself considerable freedom in prefaces such as these, and the difference between the *riwāyāt* of al-Bakkā'ī and Ibn Bukayr is due to their having attended different lectures on different occasions.

Nothing is said here about the jinn or the Arabian soothsayers. They are dealt with on fo. 35. Instead, there is an interesting summary of the attitude of the heathen Arabs towards revealed religion. "The Arabs were illiterate.[1] They did not study writing. All that they knew of heaven and hell, the resurrection, the mission of prophets and so on was the little they had heard from Jews and Christians. This teaching had no effect on their lives."

The tradition cited from 'Āṣim ibn 'Umar is worded quite differently, but it is to the same effect (cf. I.H. p. 134). That from Yaḥyā ibn 'Abdullāh to the effect that Ḥassān ibn Thābit when a child of seven or eight heard a Jew saying that the star of Aḥmad had arisen is put in this context instead of in the section dealing with the birth of the prophet (I.H. p. 102).

The story of the Jew who announced the imminent arrival of Muhammad in Medina is practically the same, though the version in the MS. is more detailed. The tale of the three associates of the Jews of Banū Qurayẓa who went over to Islam when the settlement was taken is told in somewhat different words.[2] The MS. has "they were of Hudhayl" instead of "Banū Hadl" and adds

[1] Already the word *ummī* which is applied to the prophet in the Koran is given the meaning "illiterate" instead of "gentile".

[2] The MS. points the second name Usayd. This is against the statement of S. (p. 124) that Ibn Bukayr is an authority for the form Asīd.

"they were not of Banū Qurayẓa nor al-Naḍīr: they were above that" (so T. p. 1490). At the end there is a slight addition: "Their property was in the fort with the *mushrikīn* and when the fort was taken their property was restored to them."[1] (I.K. follows Ibn Hishām with a reference to Ibn Bukayr's version in Bayḥaqī.)

The Story of Salmān the Persian

MS. fos. 23b–27a; I.H. pp. 136–42; L. pp. 95–8; I.K. pp. 311–16

This long drawn out story is told in different words. One could wish that a more important section of Ibn Bukayr's version of the *Sīra* had been preserved.

From this point onwards the MS. pursues an order of its own. Folio 27a continues with stories about the Ka'ba. Many traditions which came to Ibn Bukayr from other sources are interpolated in Ibn Isḥāq's traditions. The break in the chronology lasts until fo. 37b.

The Antiquity of the Ka'ba

MS. fos. 27a–31a; I.H. pp. 122–9; L. pp. 87–8

The greater part of this section owes nothing to Ibn Isḥāq. It is preceded by the statement that Quraysh used to venerate the Ka'ba, circumambulating it and asking for (God's) forgiveness there, the while they venerated the idols and sacrificed to them, performing the rites of the *ḥajj* and standing at the halting places.

Most of the traditions here belong to the first part of Ibn Isḥāq's work known as the *Mubtada'* and will be found in al-Azraqī's *Akhbār Makka*, pp. 31–3. One, which is a sort of midrash, is not quoted by al-Azraqī. The tradition that Muhammad took part in the *ḥajj* while he was a heathen is recorded by I.H. p. 29.

There is an interesting tradition on fo. 30a dealing with the calendar. The prophet is reported to have said "Time has come round to where it was on the day that God created heaven and earth". Ibn Abī Najīḥ explained this *ḥadīth* thus: "Quraysh used to intercalate a month in every year, and Dhūl-Ḥijja fell at the proper time only once in every twelve years. God so arranged it for his apostle that he made the pilgrimage in that month and so the apostle uttered the words in question." The author asked his

[1] It seems a little hard that the world's first monotheists should be called polytheists!

informant "What about the pilgrimage of Abū Bakr and 'Attāb b. Asīd?" He replied that they did the same as everyone else, and went on to explain that men used to make their pilgrimage in Dhūl-Ḥijja, then in the following years in al-Muḥarram, Ṣafar and so on until they went through the twelve months (and arrived again at Dhūl-Ḥijja). 'Attāb was the man whom Muhammad left in charge of the pilgrimage in the year 8 (I.H. p. 887; L. p. 597; Azr. 1, 120) in Dhūl-Qa'da.

Here instead of dating the institution of the Ḥums before or after the Elephant, the MS. says "before or after the rebuilding of the Ka'ba". Ibn Hishām's version is to be preferred. The MS. adds to the account of the restrictions which the Ḥums placed on pilgrims: "When they entered the sacred area they abandoned the provisions which they had brought and bought food from the sanctuary. They tried to obtain a garment to go round in either by hire or from charity, and if they failed they went round the Ka'ba naked." (Cf. I.H. p. 128; L. p. 87.)

The Rebuilding of the Ka'ba

MS. fos. 31b–35a; I.H. pp. 122–9; L. pp. 84–7; T. pp. 1134–9; I.K. pp. 298–305

In this section, where again the introduction differs somewhat, there are some important variants and additions where the narrative pursues the same course as the *textus receptus*. The MS. differs slightly in the story of Duwayk's part in robbing the Ka'ba. He is called Duwayl or Duwayd with a note that opinions on the point differ, and the name of Abū Ihāb has dropped out. The green stones which the workmen came upon as they demolished the old building are said to have been like spear points or teeth (*asinna*) as in T. and I.K., whereas I.H. and Bukhārī (as quoted by S.) say "humps" (*asnima*).

On fo. 32b there is a tradition belonging to the circle of stories about inscriptions at the base of the Ka'ba which is given by al-Azraqī (p. 38) with small variants, as well as the one quoted by Ibn Hishām.[1]

After the blood covenant mentioned in I.H. p. 125 the MS. contains a poem attributed to 'Ikrima ibn 'Āmir ibn Hāshim ibn

[1] There is an interesting reading in the next paragraph where, instead of *taḥāwazū* "they separated one from another" or "went their several ways" (L. p. 86) the MS. has *taḥāzabū* "they split up into factions" and this accords better with the following *wataḥālafū* "and formed alliances".

'Abdu Manāf ibn 'Abdu'l-Dār and a poem in reply by Wahb ibn 'Abdu Manāf.

There is a noteworthy difference in the status given to Abū Umayya ibn al-Mughīra ibn 'Abdullah ibn 'Umar ibn Makhzūm who is said to be lord (*sayyid*) of all Quraysh. This statement conflicts with the later theory that the family of 'Abdu'l-Muṭṭalib ought to be pre-eminent in Quraysh and we find it watered down in Ibn Hishām to the assertion that he was "the oldest that year of all Quraysh". One MS. of T. (1138) has *aysara* "richest" and in the margin *ashrafa* "noblest", and in both cases limits his priority to that particular year.

A unique and remarkable tradition is contained on fo. 33a. Ibn Isḥāq himself is the speaker. "I was sitting with Abū Ja'far Muḥammad ibn 'Alī when 'Abdu'l-Raḥmān al-A'raj, freedman of Rabī'a ibn al-Ḥārith ibn 'Abdu'l-Muṭṭalib passed us. He called him saying, 'What is this that I hear that you are saying to the effect that it was 'Abdu'l-Muṭṭalib who put the stone in its place?' 'God save you', said al-A'raj, 'I was told by a man who heard 'Umar ibn 'Abdu'l-'Azīz saying that he was told that Ḥassān ibn Thābit used to say: "I was present when the Ka'ba was rebuilt and it is as though I can see 'Abdu'l-Muṭṭalib sitting on the wall, an old man with his eyebrows held back by a turban (waiting) until the stone was brought up to him. It was he who put it in its place with his own hands."' He (Abū Ja'far) said 'Be off with you in the guidance of God'.[1] Then he turned to me and said 'This is indeed something we have never heard of. None but the apostle of God set the stone in its place with his own hands."' (The speaker continues with the accepted story of how the apostle solved the difficulty created by the rivalry of the chiefs.)

This tradition would be automatically rejected by the later professional traditionists as *munqaṭi'*, i.e. a *ḥadīth* with an incomplete *isnād*. It would also be condemned as *munkar*, "objectionable", for obvious reasons.[2]

[1] *unfur rāshidan.* See Appendix A.

[2] But incomplete attestation through a missing link in the chain of transmittors or even in some cases the anonymity of the original informant is an argument for, rather than against, the authenticity of a *ḥadīth*. The earliest authorities cared less about exact documentation of *ḥadīth*, because in their day the systematic invention of traditions about the prophet or traditions claiming his authority had not long begun. It was not until al-Shāfi'ī's influence had become paramount, when the canonical collections of *ḥadīth* were compiled, that the *isnāds* of all traditions were scrutinized and put into categories. By that time the inventors had taken care to see that their

The MS. says that the apostle was thirty-five years of age at the time, and revelation came to him about five years after the rebuilding of the Ka'ba when he was forty. He stayed in Mecca for thirteen years (after his call) and then emigrated to Medina.

When the rebuilding was completed the Ka'ba was given a roof for the first time. The MS. continues with the poem attributed to al-Zubayr. In line 7 the MS. has *masāwīna* for the common reading *musawwīna*, a variant noted by Ibn Hishām which is undoubtedly right.

Another poem by al-Zubayr and one by al-Walīd ibn al-Mughīra on the snake I have not found elsewhere. This section ends with quotations from Sūras 2, 195, and 7, 29 on the *ḥajj*.

Prophecies on the Coming Prophet and an Account of some Contemporary Monotheists

MS. fos. 35a–38b; I.H. pp. 130–50; L. pp. 90–104; I.K. as cited; S.I. p. 137

To begin with, the two versions run parallel. There is a variant in the oracle of al-Ghayṭala (where the MS. writes al-Ghayṭālija)[1] which begins: "I know whom (instead of I.H.'s "what") I know." This would make it probable that the mysterious language refers to the clan of Ka'b.[2]

There follows a tradition which I believe occurs only in S. (p. 137) in a slightly different form. It runs: "This tribe of the Anṣār used to talk about what they heard the Jews say about the apostle. The first mention of the apostle in Medina before his mission was on this wise: Fāṭima, mother of al-Nu'mān ibn 'Amr,[3]

productions were suitably documented. (See J. Schacht, *The Origins of Muhammadan Jurisprudence*, Oxford, 1950, pp. 36f.) It is impossible to suggest a motive for the invention of this story. Ḥassān ibn Thābit died in 54; 'Umar ibn 'Abdu'l-'Azīz was born in 63, nine years later, and lived until 101. A member of an intermediate generation was needed to establish contact between them, and it is unfortunate that the name of the intermediary has not been preserved.

[1] The *jīm* in the woman's name stands for the *yā* of the *nisba* ending as explained by Ibn Sikkīt, *K. al-Qalb wal-Ibdāl*, Leipzig, 1905, p. 28, so that she would be called al-Ghayṭalīya outside her own tribe. I.H. (p. 132) says that she belonged to Banū Murra. This was one of the tribes which substituted *jīm* for *yā* according to Ibn Sikkīt.

[2] In the second oracle *shu'ūb* is clearly marked with a *ḍamma*. This would conflict with my rendering (L. p. 92) but the grammatical difficulty still remains, as the MS. has the incongruous *fīhi*.

[3] From I.H. pp. 352 and 383 it is clear that this man was a Jew who was hostile to Muhammad.

brother of Banū'l-Najjār, who was a harlot during the pagan era, had a familiar. She used to relate that whenever he came to her, the house became intensely dark to those who were in it, until one day when he came to her he alighted on the wall and did not behave as he normally did. When she asked him what was the matter he said that a prophet had been sent to forbid fornication.

The form *iftaḥama* is not to be found in the lexicons, but its meaning seems clear. S. says: "Fāṭima the Najjarīya, *daughter* of al-Nu'man had a familiar from the jinn, and whenever he came to her he precipitated himself (*iqtaḥama*) upon her in her house, and when the prophet's mission had begun he came to her and sat on the wall of the house and did not enter. When she asked him why he did not come in, he said that a prophet had been sent to forbid fornication. That was the first mention of the prophet in Medina."

The tradition is interesting in that it shows that Jews were regarded as members of the Banū'l-Najjār. This was the tribe to which the prophet's maternal relations belonged. When al-Akhṭal satirized the son of Ḥassān ibn Thābit he called the Anṣār Jews.[1]

The story of Thaqīf's fear of falling stars (I.H. p. 131) is told in more archaic language, but the gist is the same. The same may be said of the next tradition. It differs in language much more widely and says nothing about soothsaying (*kihāna*) being abolished.

The grouping of traditions under the head of testimonies to the coming prophet has dictated a different order to the compiler. Ibn Bukayr cites the words of Waraqa ibn Naufal (I.H. p. 121), acknowledging Muhammad as the prophet of this people and the expected one. He breaks into verse (wisely omitted by Ibn Hishām) which is to be found in S. I, 127; I.K. II, 297; and Ibn Sa'd, I, 51, the last-named quoting five lines only.

The MS. continues with the story of the four men who eschewed idolatry and went in search of the true religion. The language is different, but the general meaning is the same, cf. I.K. pp. 237f. Of Zayd ibn 'Amr ibn Nufayl it says: "Among them there was none more upright in his affairs and intention (*sha'n*). He separated himself from idols and stood apart from the religion of the Jews and Christians and all sects except the religion of Abraham, declaring the unity of God and rejecting all else. He would not eat the sacrifices of his people and showed his hostility by separating himself from their way of life." The next tradition about him is as in I.H. p. 145. His *rajaz* verse (I.H. pp. 147f.)

[1] *Al-Aghānī* (Cairo, n.d.), XIII, 142.

appears to have suffered somewhat by faulty transmission in the MS.

There follow the two poems (reminiscent of the Koran in places) that are found in I.H. pp. 145 and 148, but with slight variants. Cf. I.K. II, 242.

The story of Zayd's persecution at the hands of his uncle and half-brother, al-Khaṭṭāb, is told in a slightly different way.

A tradition of outstanding importance follows (fos. 37b–38).[1] It is the only extant evidence of the influence of a monotheist on Muhammad by way of admonition. Ibn Isḥāq says: "I was told that the apostle of God while speaking of Zayd ibn 'Amr ibn Nufayl said, 'He was the first to blame me for worshipping idols and forbade me to do so. I had come from al-Ṭā'if with Zayd ibn Ḥāritha when I passed by Zayd ibn 'Amr on the high ground above Mecca, for Quraysh had made a public example of him (*shaharathu*) for abandoning their religion, so that he went forth from among them and (stayed) in the high ground of Mecca. I went and sat with him. I had with me a bag of meat from our sacrifices to our idols which Zayd ibn Ḥāritha was carrying, and I offered it to him. I was a young lad at the time. I said "Eat some of this food, O my uncle". He replied "Nephew, it is a part of those sacrifices of yours which you offer to your idols, isn't it?" When I answered that it was he said "If you were to ask the daughters of 'Abdu'l-Muṭṭalib they would tell you that I never eat of these sacrifices and I want nothing to do with them". Then he blamed me and those who worship idols and sacrifice to them saying "They are futile: they can do neither good nor harm", or words to that effect.' The apostle added 'After that with that knowledge I never stroked an idol of theirs nor did I sacrifice to them until God honoured me with His apostleship'."

This tradition has been expunged from Ibn Hishām's recension altogether, but there are traces of it in S. (p. 146) and Bukhārī (K. p. 63, bāb 24) where there is an imposing *isnād* going back to 'Abdullāh ibn 'Umar to the effect that the prophet met Zayd in the lower part of Baldaḥ before his apostleship. "A bag was brought to the prophet *or the prophet brought it to him* and he refused to eat of it saying 'I never eat what you sacrifice before your idols. I eat only that over which the name of God has been mentioned.' *He blamed Quraysh* for their sacrifices", etc.

Suhaylī discusses the question as to how it could be thought that God allowed Zayd to give up meat offered to idols when the

[1] See Appendix B.

apostle had the better right to such a privilege. He says that the *ḥadīth* does not say that the apostle actually ate of it; merely that Zayd refused to do so. Secondly Zayd was simply following his own opinion, and not obeying an earlier law, for the law of Abraham forbade the eating of the flesh of animals that had died, not the flesh of animals that had been sacrificed to idols.[1] Before Islam came to forbid the practice there was nothing against it, so that if the apostle did eat of such meat he did what was permissible, and if he did not, there is no difficulty. The truth is that it was neither expressly permitted nor forbidden.

I.K. (p. 239) also retains part of the original tradition which our MS. contains. He says "Zayd ibn 'Amr came to the apostle who was with Zayd ibn Ḥāritha as they were eating from a bag they had with them. When *they* invited him to eat with them he said, 'O nephew, I never eat from what has been offered to idols'."

Zayd's journey to Syria in quest of the true religion is told in slightly different words. The MS. says that the monk whom he consulted was in a church instead of "in an elevated place of al-Balqā'" as Ibn Hishām has it. The difference is probably due to a misreading rather than a different tradition. One can see what contemporary opinion of the state of Christianity was from the words ascribed to the monk which are found only here and in I.K. p. 238: "The knowledge of it (primitive Christianity) has perished and those who knew of it have passed away."

Waraqa's elegy on Zayd is shorter by three verses in the MS. and I.K. It is a fairly frequent comment of Ibn Hishām that the poems handed down by Ibn Isḥāq have suffered interpolations. The section concludes with an entirely different summary passage from Ibn Isḥāq which refers once more to the Ḥums. Cf. I.H. p. 52; L. p. 36. An addition to the text which has not come out clearly in the photograph seems to read: "The people of Najd of Muḍar used to cry Labbayka[2] towards the sacred house."

The Prophet's Call

MS. fos. 38b–40b; I.H. pp. 151–4; L. pp. 105–7; I.K. III, 26

The section opens with the famous tradition from 'Ā'isha about revelation coming to the prophet like the breaking of the dawn, but the tradition is worded differently. There is no mention

[1] Both Judaism and Christianity (Acts xv. 20 modified by I Cor. x. 23) forbade the practice. Sūra 2, 134 *et passim* prohibits Muslims from eating meat sacrificed to idols.

[2] If the word is to be read *yuhillūna.*

of visions during sleep or "true vision". It is simply said that whatever the prophet saw came like the breaking of the dawn and he remained thus as long as God willed that he should. The salutation of trees and stones is said to be the salutation of prophethood. The word used of Muhammad's devotions is *nasak*, and it is said that members of Quraysh who practised such devotions in the pagan era used to feed any of the poor who came to them. Ibn Hishām says that it was the prophet who used to do that. There is no mention of the brocade with the writing, nor of Gabriel putting pressure on the prophet, but the syntax makes it probable that the words or similar words must have been in the text once, because it continues "then he took it off me", and as the text stands the only object is the command of God. The reading *wamā 'aqra'* here is noteworthy.

The MS. contains the passage in T. p. 1150, L. p. 106 describing how Muhammad thought that he must be a *shā'ir* or a possessed person, and contemplated throwing himself from the top of a mountain. There is the graphic detail that Muhammad remained standing where he was until the daylight had almost gone. The MS. agrees with T. against I.H. in the account of the conversation between Muhammad and Khadīja and what Waraqa said.

There follow two poems attributed to Waraqa (again wisely omitted by Ibn Hishām) which are to be found in I.K. III, 11.

A tradition which is not to be found in Ibn Hishām and other authorities runs as follows: "The apostle used to suffer from the evil eye in Mecca and it came upon him swiftly before revelation came to him. Khadīja, daughter of Khuwaylid, used to send to an old woman in Mecca to charm it away. When the Koran came down to him and he suffered from the evil eye as he had before, she asked him saying 'O apostle of God, shall I not send to that old woman to charm you?' He replied 'For the present, No'."[1]

This is the last tradition from Ibn Isḥāq in Part II.

Part III

The Prophet's Call (continued)

MS. fos. 41a–43a; I.H. pp. 150–7; L. pp. 104–12; I.K. III, 23f.; T. pp. 1142f.

This section is written in a different hand. It begins by going back to I.H. p. 150 and resumes the narrative at I.H. p. 155. After the assertion that only a firm and resolute man can sustain

[1] See Appendix C.

the weight of prophecy and its responsibility there is a tradition from Ibn Munabbih to the effect that the prophet Jonah was a good man but he had a weakness in his character, so that when the weight of prophecy bore upon him he gave way like an overloaded beast, abandoned it, and fled.

The MS. continues with the touching account of the support and encouragement which Khadīja gave her husband when he was opposed by Quraysh.

The tradition about revelation coming like the breaking of the dawn already recorded on fo. 38 b is repeated word for word. The repetition perhaps is due to the second scribe having re-sorted his material without seeing that this tradition had already been recorded by his predecessor.

The story of how Khadīja proved that the visitant was Gabriel and not one of the jinn by disclosing her form when he appeared to the prophet and thus forcing Gabriel to withdraw is told in the same words.

There is a repetition of the chronology of the prophet's life which has already been given on fo. 34b: he was forty years old when revelation came; he stayed thirteen years in Mecca; and lived for ten years in Medina.

In the reference to the temporary cessation of prophecy the MS. says: "Then the prophet thought within himself from the sorrow he felt 'I fear that my companion (*ṣāḥib*) hates me and has abandoned me', to which Sūra 93 replied 'Thy Lord has not forsaken thee nor hated thee'."

The concluding words in Ibn Hishām about secrecy in communicating Gabriel's message are lacking.

The Pattern of Ritual Prayer

MS. fo. 43 a–b; I.H. pp. 157–8; L. pp. 112–14; I.K. III, 24f.; S. pp. 162f.

Here the tradition for which Ibn Isḥāq offers no *isnād* differs considerably from the pattern subsequently adopted, and the difference is afterwards discussed by I.K. on pp. 117f., where he repeats a part only of what Ibn Isḥāq wrote at this point. The MS. reads thus: "Then Gabriel came to the prophet when prayer had been ordained and he hollowed out a place in the wadi with his heel so that a spring of running[1] water (I.K. "Zamzam") gushed forth. Gabriel performed his ablution while Muhammad watched

[1] *mazn*. Possibly "rainwater" is meant.

him. He washed his face, rinsed his mouth, sniffed water up his nostrils, and wiped his head and his ears and his legs as far as the ankles, and sprinkled his pudenda. Then he arose and prayed two bowings and prostrated himself four times upon his face. Then the prophet returned, God having refreshed him, and his mind was at rest, for what he yearned for had come to him from God. He took Khadīja by the hand and brought her to the spring and performed his ablution as Gabriel had done. Then he made two bowings and four prostrations, both he and Khadīja. Thereafter the pair of them used to pray secretly."

The next tradition deals with the same subject. I cannot find it elsewhere. "Ṣāliḥ ibn Kaysān from 'Urwa ibn al-Zubayr from 'Ā'isha. When prayer was first prescribed it consisted of two bowings; then it was raised to four but remained at two for the traveller. (Ṣāliḥ) said 'I told that to 'Umar ibn 'Abdu'l-'Azīz and he said of 'Urwa "He told me that 'Ā'isha used to make four bowings during prayer while on a journey". 'Urwa came and I thought to myself (at this point the MS. is illegible). I asked him about the tradition and he related it. 'Umar said "I know nothing of these traditions of yours". Then he got up from his seat and went into his house.'"

'Alī accepts Islam

MS. fo. 43b; I.H. pp. 158–60; L. pp. 114–15; I.K. III, 24f.; T. pp. 1161f.

The MS. has the tradition given by Ibn Kathīr, and then follows the tradition from 'Afīf quoted by T. in a somewhat different form. The MS. adds that 'Afīf met al-'Abbās at Minā, and that 'Afīf sold him some goods and bought other things from him; that Muhammad came out of a tent to pray, facing the Ka'ba, and Khadīja and 'Alī did likewise. 'Afīf said that he wished that he had believed earlier and then he could have been the second Muslim. (So I.K., but T. says "the third".)

T. follows up this tradition with another which claims to rest on the authority of 'Afīf also. It comes from Ibn Isḥāq *via* Salama. It says that the prophet carried out the ritual ablutions before his prayer, and 'Afīf afterwards said that he wished that he had been the fourth.

The tradition in the MS. is suspect for two reasons: (*a*) Muhammad is said to have faced the Ka'ba, whereas I.H. p. 190 says that while he was in Mecca he faced towards Syria, i.e. Jerusalem was his *qibla*; and (*b*) it is unlikely that at this early date

Muhammad expected that his followers would conquer the Byzantines and Persians. In any case when two traditions from the same authority are at variance either or both must be wrong in whole or in part.

Abū Bakr, followed by Eight Others, Accepts Islam

MS. fo. 43b; I.H. p. 161; L. p. 115; I.K. III, 26f.; T. p. 1168

The MS. gives the longer account as it stands in I.K., followed by the prophet's words about Abū Bakr's promptitude in accepting Islam. (The word used is *'attama* instead of the awkward *'akama*.) The first to follow the apostle was his wife Khadīja; the first male was 'Alī at the age of ten; the third Zayd ibn Ḥāritha; and then Abū Bakr.

A Reference to Muhammad in Scripture and the Epithets which He applied to Himself

MS. fo. 44b

Umm al-Dardā' said that she asked Ka'b al-Ḥibr what reference he found to the apostle in the Old Testament (*Taurāt*). He answered: "We find Muhammad the apostle of God. His name is al-Mutawwakil. He is not harsh or rough; nor does he walk proudly in the streets. He is given the keys[1] that by him God may make blind eyes see, and deaf ears hear, and set straight crooked tongues so that they bear witness that there is no god but Allah alone without associate. He will help and defend the oppressed." This would seem to be a garbled version of Isaiah xlii. 2–7.

Another tradition is that the apostle was heard to say that he had five names: Muhammad and Aḥmad and al-Māḥī by whom God expunged disbelief; al-'Āqib (the last one); and al-Ḥāshir by whom God would assemble men before his feet.

The Muhājirūn accept Islam. Quraysh show Resentment

MS. fos. 44b–49a; I.H. pp. 162–71; L. pp. 115–21; I.K. pp. 39–49; T. pp. 1171f.

There are slight variants in the names of the Muhājirūn which may be of interest to historians. The concluding paragraph about the hostility which the conversion of these men caused among Quraysh is somewhat different.

The MS. contains the story of how the prophet entertained the

[1] Possibly we should read *maṣābīḥ* "lamps" for *mafātīḥ*.

sons of 'Abdu'l-Muṭṭalib to a meal on two successive days in an endeavour to get them to listen to his message.[1] The parallel passage in T. p. 1172 contains an obvious piece of 'Alid propaganda which the MS. lacks. In T. 'Alī was the only one to offer to help Muhammad, who appointed him his executor (*waṣī*) and successor (*khalīfa*).

The MS. continues as in I.H. pp. 166–70, but the order of the narrative differs from Ibn Hishām's recension. To the story of Abū Ṭālib's staunchness in summoning the Banū Hāshim and Banū'l-Muṭṭalib to defend their kinsman and the defection of Abū Lahab, the MS. adds that the latter stirred up the Banū Hāshim against Muhammad, and that Banū'l-Muṭṭalib were in an alliance which excluded the Banū 'Abdu Manāf.[2] It also cites two poems attributed to Abū Ṭālib which I do not find elsewhere. It continues with Abū Ṭālib's poem on I.H. p. 171 from which it omits the last line.

A new note is struck when Abū Ṭālib attacks Abū Lahab in five lines of verse not to be found elsewhere. They are prefaced by a note that Abū Lahab's mother was from Khuzā'a while 'Abdullāh, the prophet's father, and al-Zubayr and Abū Ṭālib himself, were the children of Fāṭima, daughter of 'Amr, who belonged to the clan of Makhzūm. In I.H. pp. 70 and 113 Abū Lahab's mother's name is given as Lubnā. In the MS. it is said that she was called Samājīj which could mean simply "ill-favoured" or perhaps "without any good quality". (The final *jīm* has been explained on p. 25.)

In another poem of six lines Abū Ṭālib warns Quraysh of the dangers of internecine war. I cannot find this in any other source. Then follows the story of how al-Walīd ibn al-Mughīra gathered Quraysh together to decide how they were to speak of Muhammad when the Arabs gathered at the local fair. Then the attempt of Quraysh to buy off Abū Ṭālib by giving him 'Umāra ibn al-Walīd to adopt as a son in place of Muhammad. The order of the MS. seems to be better than that of I.H. There are a few textual differences: Abū Ṭālib says "Don't you know that if a camel has lost its foal she has no feeling for any other?"

[1] Incidentally the MS. shows that the correct word in T. p. 1172 is *lahadda mā* "how greatly" (has your host perverted you!). The Leyden text *laqidman* seemed to me to be hopeless and I left it untranslated in L. p. 118 because the context did not imperatively demand another word.

[2] Cf. I.H. p. 169, ll. 4 and 6 of the poem, L. p. 120; I.H. p. 175, l. 57 of the poem, L. p. 125; I.H. p. 176, l. 65 of the poem, L. p. 126; I.H. p. 245, l. 7 of the poem, L. p. 170; W. M. Watt, *Muhammad at Mecca* (Oxford, 1953), p. 6.

The Persecution of the Apostle's Companions

MS. fos. 48a–49a; I.H. pp. 168f.; L. pp. 119f.; I.K. III, 42–57 and 86

There are textual differences in the account of Abū Ṭālib's interview with his nephew. The inevitable poem (found in I.K. p. 42) follows.

The next passage is of importance because of the light it throws on the growth of the poetry contained in the *Sīra*. Ibn Bukayr gives us only the first seven lines of a poem which in I.H. has no less than ninety-four lines. But Ibn Hishām himself says at the end "This is as much as I regard as authentic, but some authorities on poetry object to most of it". From this it may be regarded as certain that in Ibn Hishām's day there was even more of the poem in circulation. The introduction to these verses in the two recensions shows that considerable editorial effort has been spent upon this poem. Ibn Kathīr bestows extravagant praise on it. He says that it is of such excellence that none but the man to whom it is ascribed could have composed it. It is finer even than the Seven Mu'allaqāt! Al-Umawī, he says, has given it in full with even more verses in his *Maghāzī*. He himself gives ninety-five verses.

The Deed of Boycott

MS. fos. 49a–52a; I.H. pp. 230–2 and 249f.; L. pp. 159–61, 173f.; I.K. III, 86f.

The narrative runs as in I.K., departing both from him and I.H. in reference to Abū Lahab's boast that al-Lāt and al-'Uzzā had won. A *fakhr* poem attributed to Ṣafīya daughter of 'Abdu'l-Muṭṭalib follows. I have not been able to find this elsewhere. This is followed by the poem of Abū Ṭālib which has eleven lines as against I.H. and I.K. who give fourteen, and by another of fifteen lines which appears only here.

The next passage seems to be unique. It describes how Abū Ṭālib took his company to the Ka'ba and invited God's vengeance on those who had severed the ties of kinship and resolved to shed their blood. Then he went to the gathering of Quraysh and threatened them with divine vengeance. They replied that there could be no peace between him and them unless he surrendered "this foolish young man" so that they might put him to death. Thereupon Abū Ṭālib withdrew with his following to the narrow quarter of Mecca.

Evidently the proper sequence of events has been disturbed, for the passage continues with the note that when 'Umar and 'Amr ibn al-ʿĀṣ and ʿAbdullāh ibn Abī Rabīʿa came back to report the failure of their mission to the Negus (I.H. pp. 217–21) the Quraysh were still more incensed against Muhammad and his companions. This mention of ʿUmar is extremely interesting. I.H. mentions only the two last named and expressly asserts that Quraysh sent *two* men. Presumably only one man could be referred to simply as ʿUmar without further qualification—the famous companion and caliph.

The MS. goes on to describe how Quraysh kept the Muslims in a state of siege and beat them if they met them in the streets. They cut off their supplies in the markets and would not allow anyone to bring them food. They even hurried to the merchants before them and bought up everything so that prices rose against the Muslims. Al-Walīd ibn al-Mughīra used to send a crier round to tell people to buy food from any man who had it and so raise the price against the Muslims. If they had no ready money they were to buy the food on credit. This state of affairs lasted for some three years until most of the Quraysh were perturbed by the sufferings of the Banū Hāshim and some of them wanted to dissociate themselves from the boycott. Abū Ṭālib was so anxious for the safety of Muhammad that he used to put him between his sons and himself when he lay down at night lest he should be killed. There follows a poem in which Abū Ṭālib laments the sufferings of these days. It is not recorded elsewhere. The narrative here is more vivid than that in I.H.

The story of Abū'l-Bakhtarī's striking Abū Jahl with the shank (not the jawbone as in I.H.) of a camel is followed by three lines of indifferent verse. The incident of the eating of the deed by worms is told in a somewhat different way. Abū Ṭālib conceals the news which Muhammad gave him from God, and when at last he comes to divulge it the Quraysh think that he has been forced by starvation to surrender Muhammad to them. The narrative then follows pretty much the same course as I.H. p. 249. In place of the poem there given there are two others of seventeen and eleven lines respectively.

Abū Ṭālib gives Protection to Abū Salama

MS. fos. 52a–b; I.H. pp. 244–5; L. p. 170; I.K. III, 93

This story is told in different words. It is important to notice that the poem attributed to Abū Ṭālib has only five lines instead of the nine given by I.H. who says that he has omitted one.

The Destruction of the Deed of Boycott

MS. fos. 52b–53a; I.H. pp. 247–50; L. pp. 172–3; I.K. III, 95–8

The differences here are slight. Once again Abū Ṭālib's poem is reduced: this time from twenty-six to six lines.

'Umāra ibn al-Walīd ibn al-Mughīra's Assault on 'Amr ibn al-'Āṣ and his Wife and the Way 'Amr took Vengeance

MS. fos. 53a–54b; I.K. III, 76

This long story, which is reminiscent of the *Arabian Nights*, is found only here and in *Al-Aghānī*[1] (with variations) on al-Wāqidī's authority, but a garbled form of it is given by Ibn Kathīr who evidently knew the version of Ibn Bukayr from Ibn Isḥāq. See also Abū Nu'aym.[2] It has but little to do with the biography of the prophet and therefore Ibn Hishām in accordance with his principle of omitting all extraneous matter has dropped it from his edition of the *Sīra*. The following summary has been somewhat bowdlerized. The story runs that after his visit to Abū Ṭālib with men of Quraysh, 'Umāra accompanied by 'Amr ibn al-'Āṣ set sail for Abyssinia on a trading venture. 'Umāra was a notorious and handsome Don Juan and on the voyage, flushed with wine, he made violent overtures to 'Amr's wife who had considerable difficulty in repelling them. Waiting his opportunity he took 'Amr at a disadvantage and pushed him into the sea; but 'Amr was a good swimmer and regained the deck. 'Amr realized that 'Umāra had intended to murder him, and so he waited until opportunity came to encompass his ruin. He took the precaution of writing to his father in Mecca to declare that he would not be responsible for any action against 'Umāra, and his father did likewise. The Banū Makhzūm and Banū'l-Mughīra made a similar declaration and so 'Amr's course was clear: he could bring about 'Umāra's death without involving his family in a vendetta.

'Umāra resumed his pastime of pursuing other men's wives as soon as he arrived in Abyssinia and within a short while was able to boast to 'Amr that the wife of the Negus was his latest conquest. 'Amr artfully refused to credit the story until he could bring proof, although he knew full well that 'Umāra's boast was well founded. The vainglorious fellow fell into the trap and

[1] Cairo (1936), IX, 55 f.
[2] *Dalā 'il al-Nubūwa*, pp. 205 f.

agreed to coax the Negus's wife to anoint him with the perfume that the Negus kept for his personal use and allowed to none other. The woman also fell into the trap and gave him a pot of the ointment which 'Umāra handed to 'Amr. After a suitable interval 'Amr went to the Negus and reported the matter to him. The Negus immediately recognized the smell of the ointment and ordered some women who are called witches (*sawāḥir*) to strip 'Umāra and inflate his member. The wretched creature was then allowed to go, and he fled to the wilderness where he lived with the beasts until the caliphate of 'Umar. At that time some of the Banū'l-Mughīra among whom was 'Abdullāh ibn Abū Rabī'a ibn al-Mughīra whose pre-Islamic name was Bujayr were in Abyssinia. They lay in wait for 'Umāra by a pool where he came down to drink with the wild asses. He smelt the presence of human beings and ran away until, exhausted by thirst, he had to return to the water. 'Abdullāh asserted that he overtook him as he fled once more after drinking his fill and seized him. 'Umāra besought him to let him go for he would die if he were held, but 'Abdullāh kept hold of him and he died on the spot. He buried him there and reported that the whole of his body was covered with hair. The story, which may well be founded on fact, is not improved by some seven lines of doggerel attributed to 'Amr.

The Conversion of Ḥamza

MS. fos. 54b–55b; I.H. pp. 184–5; L. pp. 131–2; T. pp. 1187–8; I.K. III, 33

The first half of this section agrees fairly closely with the other three authorities: the second half is found only in I.K.

The MS. adds a few details: the woman concerned lived above al-Ṣafā; Ḥamza was a heathen at the time he championed Muhammad; and Ḥamza made a much stronger confession of his faith as a Muslim when Banū Makhzūm accused him of changing his religion. Six lines of *rajaz* which Ḥamza aimed at Abū Jahl are not to be found elsewhere.

The second half contains what looks like a continuation of the story in the *Sīra*. Ḥamza returned to his house where the Satan came to him and told him that it would have been better had he died than followed this Ṣābi' and abandoned the religion of his fathers. Ḥamza prayed to God to show him the right way and spent a sleepless night in doubt. In the morning he went to the apostle and told him of his troubled mind. Muhammad explained

Islam to him and he accepted it and begged him to proclaim it openly. Then follow nine verses which have been quoted by S. p. 186.

A List of Those who Migrated to Abyssinia

MS. fos. 55b–56b; I.H. pp. 208–15; L. pp. 146–8; I.K. pp. 66–9

The order of the lists differs widely from I.H.

The Verse Inserted into the Koran at the Instigation of Satan

MS. fo. 56b; T. pp. 1192f.; S. p. 229; I.H. p. 241; L. p. 165

There can be little doubt that Ibn Hishām cut out some of the text which came to him because he gives no reason for the sudden conversion of the people of Mecca and leaves it unexplained. The full story hitherto has been known only from Ṭabarī who quoted Ibn Isḥāq on the authority of Salama. In that version it is made clear that it was the prophet's desire to end the estrangement between him and his people and to make it easier for them to accept Islam that prompted him to yield to the suggestion of Satan and add the words "These are the exalted cranes (*gharānīq*) whose intercession is to be hoped for" (or, in another version, "approved").

The MS. agrees with Salama's report from Ibn Isḥāq that the emigrants returned from Abyssinia because they heard of the conversion of Quraysh in consequence of the concession to polytheism, but strangely enough it does not quote the offending words. Presumably they were deliberately omitted and readers must have known what they were because otherwise the narrative would be unintelligible. Two verses are referred to, but the second is not quoted. In view of its interest I give a translation of the MS.: "(The emigrants) remained where they were until they heard that the people of Mecca had accepted Islam and prostrated themselves. That was because the chapter of The Star (53) had been sent down to Muhammad and the apostle recited it. Both Muslim and polytheist listened to it silently until he reached his words 'Have you seen (or, "considered") al-Lāt and al-'Uzzā?' They gave ear to him attentively while the faithful believed (their prophet). Some apostatized when they heard the *saj'* of the Satan and said 'By Allah we will serve them (the *Gharānīq*) so that they may bring us near to Allah'. The Satan taught these two verses to every polytheist and their tongues took to them easily. This weighed heavily upon the apostle until Gabriel came to him and complained to him of these two verses and the effect that they had upon the people. Gabriel declined responsibility for them and said 'You recited to the people something which I did not bring you from God and you said what you were not told to say'. The apostle was deeply grieved and afraid. Then God sent down by way of comfort to him: 'Never did we send an apostle or a prophet before you but when he wished Satan cast a suggestion into his wish' as far as the words 'Knowing, Wise'" (Sūra 22, 51).[1]

Ibn Kathīr gives a fantastic reason for the conversion of the Meccans and says that Ibn Isḥāq's tradition is not sound. He says that he himself has not quoted the story of the *gharānīq* because there might be some who heard it for the first time and would not be able to take a right view of it.

Suhaylī with his customary honesty makes no bones about it. He says that the cause of the return of the emigrants was as we have heard, and he also tells us that besides Ibn Isḥāq, Mūsā ibn 'Uqba handed on the tradition. He says that traditionists reject this *ḥadīth*, and those who accept it have ways of explaining it. One of these, he says, namely that Satan spoke the words which were broadcast through the town but the apostle did not utter them, would be excellent were it not for the fact that the tradition asserts that Gabriel said to Muhammad "I did not bring you this".

Some of the Refugees return from Abyssinia

MS. fos. 56b–57a; I.H. p. 241; L. pp. 167–9; T. p. 1194; I.K. III, 91

The MS. says nothing about the repudiation of these verses and goes on to show that the refugees were unaware that they had been disavowed and so returned to Mecca expecting to find that their townsmen had become Muslims. As soon as they reached the neighbourhood of Mecca they learned that the report was false. They could not face the journey back to Abyssinia and yet they dared not enter the town without a guarantee of safety. They stayed outside until everyone of them found someone who was willing to take him under his protection.

The wording of the MS. differs widely from I.H. and T.

[1] See Nöldeke, *Geschichte des Korans*, I, 100f. for a list of the Arab authorities who quote or discuss this interjection into the text of the Koran. To this should now be added a monograph on the subject by Ibrāhīm al-Kurānī which I wrote in *B.S.O.A.S.* (1957), XX, 291–303.

'Uthmān ibn Maẓ'ūn resigns al-Walīd's Protection

MS. fos. 57a–b; I.H. pp. 243–4; L. pp. 169–70; I.K. III, 92

On the whole the MS. preserves a better text. There is nothing new except an indifferent poem of seven lines put into the mouth of 'Uthmān.

'Umar becomes a Muslim

MS. fos. 57b–59a; I.H. pp. 224–30; L. pp. 155–9; I.K. III, 79–82

The MS. tells us a great deal about the circumstances of 'Umar's conversion. It begins at the statement that 'Umar became a Muslim after the apostle's companions had gone to Abyssinia and continues with the story of 'Umar's courtesy to the wife of 'Āmir ibn Rabī'a whose name is given as Laylā. The tradition from her is much shorter and the language is very different. A comparison of the *riwāyas* of Yūnus and Bakkā'ī shows that the language of tradition was not thought to be of much importance provided the general sense was conveyed.

The MS. adds to the story of 'Umar's visit to the apostle when he met Nu'aym on the way, the detail that Quraysh had sent him to kill Muhammad. The language again differs considerably from that of I.H. A violent dispute broke out between the two men and 'Umar accused Nu'aym of being a Ṣābi'. From this point the MS. diverges from the known text. It adds that it was the apostle's custom to commend the poor to the charity of the well-to-do and Khabbāb ibn Aratt was put into the care of 'Umar's cousin and brother-in-law Sa'īd ibn Zayd ibn 'Amr ibn Nufayl. The Sūra Ṭā Hā had been sent down and the apostle had prayed on the Thursday night that 'Umar or Abū'l-Ḥakam would be converted. 'Umar's cousin and his wife said "We were hoping that the apostle's prayer for 'Umar would be answered and it was".

'Umar came to the door as Khabbāb was studying under her guidance the Sūra Ṭā Hā and also "When the sun is overthrown" (Sūra 81, 1). The polytheists used to call this reading[1] "rubbish". When 'Umar came in his sister saw that he meant mischief and hid the sheet from which they were reading. Khabbāb slipped away into the house. 'Umar asked what was the gibberish he had heard, to which she answered that it was merely conversation between them. 'Umar upraided her and swore that he would not leave the house until he knew what had been going on; where-

[1] Literally "study".

upon Sa'īd said "You cannot get people to agree with your vain desires while the truth lies elsewhere, 'Umar". This so enraged 'Umar that he attacked Sa'īd and knocked him down. The story continues on the general lines of I.H. p. 226 but in different words. It is again implied that the *ṣaḥīfa* contained Sūras 20 and 81 and that 'Umar read them. S. 1, 217 knew of this tradition from Yūnus, but the form in which he quoted it looks like a conflation of I.H. and Yūnus. The MS. goes on to say that 'Umar asked what Islam implied and when he was told that he must testify to the One God and the apostleship of Muhammad and renounce all rivals to God and disbelieve in al-Lāt and al-'Uzzā, he at once did so. Khabbāb then appeared from within the house and congratulated 'Umar because God had answered the apostle's prayer on his behalf. 'Umar then set out for the apostle's abode at the foot of al-Ṣafā. The apostle had heard that he was seeking him in order to kill him, but he had not been told of his conversion. The apostle (not Ḥamza as in I.H. in a somewhat different sense) said "If God wishes 'Umar well he will follow Islam and believe the apostle; if not it will not be easy for us to kill him". When 'Umar was admitted the apostle seized him by his garments (there is no mention of violence as in I.H.) and said that he did not think he would end his evil way until God brought down his wrath upon him as he had upon al-Walīd ibn al-Mughīra. He prayed that God might guide 'Umar. 'Umar laughed as he pronounced the *shahāda*. The Muslims, who numbered over forty men and eleven women, praised God so loudly that those outside their quarter heard them. 'Umar then gave vent to seven verses which are quoted by S. (p. 218).

'Umar then called for action: Let them proclaim God's religion in Mecca and if the people show hostility let them make an end of them: if they behave fairly let them accept the situation. He then went forth with his companions and sat in the mosque. The Meccans were dismayed when they saw that 'Umar had become a Muslim. (I cannot find this paragraph elsewhere.)

Then follows the tradition from Nāfi' (I.H. p. 229) with a few verbal changes.

A tradition from Munkadir is not to be found elsewhere. A Badū asked what had become of the tall, overbearing fellow with a bald forehead, for he would be a powerful force for good or evil in the near future. He was referring to 'Umar.

The First Public Recitation of the Koran in Mecca

MS. fo. 59b; I.H. p. 202; L. pp. 141–2; T. p. 1188

The differences here are trifling. This is the end of Part III in the MS.

Part IV

Quraysh listen to the Apostle Reciting the Koran

MS. fo. 60a; I.H. pp. 203–4; L. pp. 142–3

Essentially the same. There is a variant reading of interest to philologists: *tajāthaynā* is written for *tajādhaynā* in I.H. No doubt S. knew of this reading when he said (p. 201) that often *al-jādhī* and *al-jāthī* are used indifferently.

The Persecution of the Muslims and their Sufferings

MS. fos. 60a–62a; I.H. pp. 205 f.; L. pp. 143–5; I.K. p. 57

The MS. omits the passage in I.H. p. 204 which is to all intents and purposes a commentary on Sūra 17, 47 f., and it cuts down the report in I.H. p. 205 to the mere statement that Quraysh persecuted the Muslims. It implies that the punishment of Bilāl endangered his life but gives no details. It preserves a poem of six lines not to be found elsewhere in which 'Ammār ibn Yāsir mentions Bilāl and his companions and their sufferings and how Abū Bakr gave them their freedom. Then follows Abū Quḥāfa's advice to Abū Bakr to free some sturdy fellows rather than the weak creatures he had redeemed from slavery, with the same quotation from the Koran. The story of the death of Sumayya, mother of 'Ammār, under her sufferings is told in different words.

The MS. adds that Yāsir was a slave of Banū Bakr having been sold to Banū'l-Ashja' ibn Layth. They married him to Sumayya who gave birth to 'Ammār. Sumayya was a slave woman and they freed her with 'Ammār and Yāsir. The MS. then passes to the *ḥadīth* of Ḥakīm, I.H. p. 207.

Then follow four new additions to the *Sīra*. The first is from Ṣāliḥ ibn Kaysān from one of the family of Sa'd ibn Abū Waqqāṣ. A certain Muṣ'ab ibn 'Umayr, who was afterwards killed at Uḥud, was a young man who lived a comfortable life in Mecca until he became a Muslim. After a time he became so emaciated by his privations that he had to be picked up and carried about by his friends.

The second deals with the same man and comes ultimately from Muḥammad ibn Ka'b the Quraẓite on the authority of Yazīd ibn Ziyād. 'Alī was heard to say that they were sitting in the mosque with the apostle when suddenly Muṣ'ab appeared clad only in a garment patched with skin. The apostle wept at seeing the contrast between a once opulent person and the miserable object now before him. He asked how they would like to be clad in fine clothes and have plenty of food placed before them and cover the walls of their houses as they covered the Ka'ba; to which they answered that it would be a welcome change for the better. for they could devote themselves to God's service and would have enough to eat. The apostle replied that things were better as they were.

In describing the straits to which the beleaguered Muslims were reduced the next story does not spare the fastidious. The speaker is the man concerned. He says that he went out one night to urinate, and hearing a splashing noise he looked to see what caused it, and found a piece of camel skin. He took it home and washed it, heated it, crushed it between two stones, and then ate it dry as it was, and drank some water to wash it down. This gave him strength for three days.

The fourth story is told by Muḥammad ibn Ka'b, the authority for the second story above. He quotes the words of 'Alī who says that on a winter's day he left the apostle's house in which there was no food, ravenous with hunger, hoping to find something to eat. He passed by a Jew who was watering his land and went up to look at him through a gap in the wall. "Ho, Arab," said the Jew, "would you like to empty the bucket at a date a time?" 'Alī accepted the offer, and when he had earned a handful of dates he let go of the bucket and ate them. Then, after a drink of water, he went to the mosque where he found the apostle.[1]

So far as I know this is the only evidence that Jews owned land in Mecca, and it might be thought that the mention of a mosque (*masjid*) and the fact that the narrator was a man of Qurayẓa must indicate that the story belongs to Medina rather than to Mecca. But (*a*) there is more than one mention of a mosque (a place of worship) in Mecca before Islam was established there, cf. I.H. pp. 185 and 233 *et passim*; (*b*) this same Muḥammad ibn Ka'b is responsible for the Meccan tradition in I.H. p. 185; and (*c*) the

[1] See Appendix D.

context in which this tradition about the Jewish landowner is found is devoted to the sufferings of the Muslims in *Mecca*. Doubtless the story has been dropped by Ibn Hishām as being disrespectful to the memory of one of the prophet's greatest companions.

Another tradition which is not in the *Sīra* follows: 'Umar is reported to have said that he went in to see the apostle and found him lying on a mat, part of his body on the ground, his head supported by a leather pillow stuffed with palm leaves, while above his head an evil-smelling hide hung from the roof of the upper room in the corner of which were a few leeks.

Muhammad forces Abū Jahl to pay his Debt to the Irāshite

MS. fos. 62a–b; I.H. pp. 257–8; L. pp. 177–8

Here there are only trifling differences.

Controversies with the Polytheists and Rabbis

MS. fos. 62b–64b; I.H. pp. 187–97; L. pp. 133–9; cf. I.K. III, 50f. and 63f.

There is no variant of any importance in this long passage except perhaps that the tradition that eight verses of the Koran came down with reference to al-Naḍr's claim to rival Muhammad in story telling is furnished with a slightly longer *isnād*, albeit inexact, in which Ibn Isḥāq says that he was told the tradition by an old man of Mecca more than forty years before. I.K. p. 52 says that the old man came from Egypt. The long extract from Sūra 17 in I.H. is reduced to a single verse in the MS.

What 'Utba said about the Prophet

MS. fos. 65b–66a; I.H. pp. 185–7; L. pp. 132–3; I.K. III, 63

The MS. adds to 'Utba's words "Perhaps that which the familiar spirit brings is poetry which burns within your breast, for you, O Sons of 'Abdu'l-Muṭṭalib, can do what others cannot".

The passage ends with a poem ascribed to Abū Ṭālib said to be in praise of 'Utba when he replied to Abū Jahl saying "We do not deny that Muhammad may be a prophet".

MS. fos. 66b–67b

Here follow three isolated traditions which have no place in I.H. In the first the prophet promises that if Quraysh follow him in the true religion their power and wealth will be multiplied. Quraysh reply in the words of Sūra 28, 57.

The second, resting on the authority of Ibn 'Abbās, says that the verse about "the accursed tree" came down in reference to Abū Jahl. This verse is twice cited in a different context in I.H. (pp. 240 and 265).

The third introduces a poem of twelve lines (not found elsewhere) thus: "It is alleged that after his conversion to Islam 'Umar made mention of what Quraysh had seen by way of warning when Abū Jahl planned evil against the apostle. (The reference is to the story in I.H. p. 190.) Some say that Abū Ṭālib spoke the lines. God only knows who the author was." There could hardly be a more barefaced confession of forgery. Had either of these men so closely associated with the prophet composed these lines they would have been known as their work and have been preserved in honour. As they stand it is admittedly not known whether they were the work of a Muslim or a man who never renounced the heathenism of his fathers. This is a typical example of the work of the forger which runs throughout the *Sīra*. He has used the story of Abū Jahl taking up a stone to kill the prophet and woven round it his wretched doggerel.

The Migration to Abyssinia

MS. fos. 67b–71a; I.H. pp. 208–23; L. pp. 146–54; I.K. III, 66–79

The heading in the MS. is "The *First* Migration to Abyssinia", but the narrative seems to be mainly concerned with a later exodus. The introduction begins: "...It was the last testing (of their steadfastness in Islam) which drove out the Muslims who emigrated after those who had gone before to Abyssinia" (*v.s.*). There follows the tradition from al-Zuhrī going back to Umm Salama the prophet's wife to the effect that the prophet's companions were being oppressed and seduced from their religion; the apostle was protected by his uncle and his clan so that he did not suffer anything unpleasant himself, though he was unable to help his companions. Therefore he told them to go to Abyssinia. They went in batches until they formed a community in comfortable quarters where they could practise their religion without fear of injustice. The MS. continues as in I.H. p. 217, but in quite different words, to describe the mission of 'Amr and 'Abdullāh to the Negus to persuade him to send back the refugees. The text is given in I.K. The story of the struggles of the Negus with his enemies is told in different words, again to the same effect. Cf. I.K.

Two traditions (the first in I.K. p. 77) follow: the first to the effect that the only person who actually spoke to the Negus was 'Uthmān ibn 'Affān. The second explicitly denies this and says that only Ja'far spoke to him.

An interesting tradition which I cannot find elsewhere is then given. The young Abyssinians were keenly appreciative of the beauty of the prophet's daughter Ruqayya who was married to 'Uthmān. They used to gaze at her and dance up to her whenever they saw her, so that their public admiration of her charms annoyed her.[1] Howbeit they were careful not to annoy any of them because of their consideration for foreigners. When the Negus marched against his enemy the Muhājirūn marched with them, and God slew the enemy to a man.

At this point the story of the arrival in Mecca of about twenty Christians who had heard of the prophet from reports from Abyssinia is interpolated. It is practically the same as I.H. p. 259.

The MS. then resumes with a tradition resting on the authority of Abū Hurayra that the prophet took out his companions and ranged them in ranks behind him. He cried *Allāhu Akbar!* four times, and when he was asked why he had done this he said that it was for the Negus who had died that day (so Bukhārī, XXIII, bāb 65).

There follows on 'Ā'isha's authority the tradition that a light was continually seen over the Negus's grave as in I.H. Ibn Isḥāq states that the name of the Negus was Maṣḥama which in Arabic would be 'Atīya (gift) and that Negus is a dynastic name like Chosroes and Hercules. Bukhārī (*loc. cit.*) and others state that his name was Aṣḥama. I.K. p. 77 says that in a MS. which al-Bayḥaqī had corrected, the form was Aṣḥam.

The MS. continues: "My father Isḥāq said, 'I saw Abū Nayzar, the Negus's son. Never have I seen a man, Arab or foreigner, larger, taller, or more handsome than he. 'Alī came on him in the company of a merchant in Mecca and bought him and gave him his liberty as some return to the Negus for his kindness to Ja'far and his companions. I asked my father if [Abū] Nayzar was a black man like the Abyssinians and he replied that had I seen him I should have said that he was an Arab.'"

[1] The writer uses the word *darkala* to describe the way in which the Abyssinians danced up to her. Arab lexicographers recognize this as an Abyssinian word.

Another apparently unknown tradition on the authority of Fāṭima, daughter of al-Ḥusayn, is that some Abyssinians came to Abū Nayzar after 'Alī had given him his freedom, and 'Alī entertained them for a month. They told him that their country was in a troubled state, and invited him to return with them so that they might make him their king. He replied that he could not accept their invitation because God had honoured him with Islam. They were disappointed and returned to their own country. She continued: "What a fine man (Abū Nayzar) was except that he was addicted to strong drink!"

The MS. then quotes two poems ascribed to 'Abdullāh ibn al-Ḥārith as in I.H. pp. 215 f.; the poem ascribed to Abū Ṭālib, I.H. p. 217; and a second poem of four lines by the latter which is not to be found elsewhere.

The section ends with a story that a party headed by 'Uthmān saw 'Abdu'l-Raḥmān ibn 'Auf, and 'Uthmān exclaimed that none could surpass him in merit for he had taken part in both hijras: one to Abyssinia and the other to Medina.

The Names of Those who Migrated to Abyssinia

MS. fos. 71b–73a; I.H. pp. 208–16; L. 146–50; I.K. III, 66–9 and 83

The names of the first ten agree but there are a few discrepancies in the genealogies and some notes on the subsequent fate of some worthies not to be found in I.H. at this point.

As in I.H., Ja'far heads the second list which is cast in a different order altogether. Again there are some discrepancies. I.H. gives the total as eighty-three, while the MS. says "just over eighty". There is doubt about one or two individuals. The tradition about Ja'far's last fight (I.H. p. 794) is given in slightly different words.

Here is inserted a poem of two lines ascribed to Abū Sa'īd ibn al-'Āṣ which does not appear to exist elsewhere. This is followed by a lampoon on Abū Ḥudhayfa from Hind, daughter of 'Utba, which this MS. preserves. Then comes the text of a short letter from Muhammad to the Negus summoning him to accept Islam and quoting Sūra 3, 57.

The poem of three lines attributed to 'Abdullāh ibn al-Ḥārith, found in I.H. p. 216, contains a valuable correction to the last word: *thaghr* for the difficult *naqr*.

How Quraysh annoyed the Apostle

MS. fo. 73b; I.H. pp. 183–4; L. pp. 130–1; T. p. 1185; I.K. p. 46

Identical but for one word.

The Prophet offers himself to the Bedouin

MS. fo. 74 a–b; I.H. p. 281; L. pp. 194–7; I.K. p. 138

Ibn Isḥāq's introduction to this section is shorter in the MS. The name of the chief of Kinda is given as Fulayḥ against all other authorities which have Mulayḥ. The tradition in I.H. from Muḥammad ibn 'Abdu'l-Raḥmān is given in the MS. as from al-Zuhrī in continuation of the preceding tradition.

There follows a short anonymous tradition to the effect that when Abū Sufyān heard that Khufāf ibn Aymā' ibn Raḥada had become a Muslim he said "The chief of Banū Kināna has changed his religion tonight". (This man is mentioned in I.H. pp. 440 and 927.)

A tradition from Sālim ibn 'Abdullāh ibn 'Umar, only found here, runs: "A man from Quraysh in Mecca came to the apostle and said 'Is it true, Muhammad, that you forbid holding captives?', meaning captured Arabs. The apostle replied 'Certainly'. Thereupon the man turned round and thrust his bare behind into the apostle's face. The apostle cursed him and invoked God against him, and God sent down 'It is nothing to do with you whether He pardons or punishes them. They are a sinful people' (Sūra 3, 123). After that the man became a firm Muslim."

Muhammad's Foster-father accepts Islam

MS. fo. 75a

Ibn Isḥāq says that his father told him on the authority of some men from Banū Sa'īd ibn Bakr that when the apostle's foster-father al-Ḥārith ibn 'Abdul-'Uzzā came to Mecca the Quraysh told him that Muhammad preached that the wicked would go to hell and the obedient to heaven, and that he had disrupted their community. He went to the apostle and asked him about the matter and was so kindly received that he accepted Islam forthwith.

Abū Bakr is taken under the Protection of Ibnu'l-Dughunna

MS. fo. 75 a–b; I.H. pp. 245–6; L. pp. 171–2

The MS., which, like I.H., purports to quote 'Ā'isha's words, arranges the discourse quite differently though the general sense is the same.

The Death of Abū Ṭālib

MS. fos. 75 b–77 a; I.H. pp. 277–9; L. pp. 191–2; I.K. p. 122

Though no new facts are brought to light this account differs widely from I.H. and I.K. The tradition on the authority of Ibn 'Abbās is reported after a longer one dealing with the same subject. This is substantially the same as I.H. The passage ends with a lament of fourteen lines on Abū Ṭālib ascribed to his son 'Alī. This is not found elsewhere. Here Part IV ends.

Part V

The Death of Khadīja

MS. fo. 78 a–b; I.H. p. 277, cf. pp. 121 and 1001; L. p. 191; I.K. v, 293 f.; S. II, 366

The first tradition agrees with I.H., except that it says that the apostle used to "confide in" Khadīja instead of "complain to her of his troubles". Yūnus ibn Bukayr continues "all that is said here about the prophet's wives was dictated by Ibn Isḥāq word for word".

For some unknown reason I.H. gives a version of the second tradition in his own name. Ibn Isḥāq's version runs thus: "The first woman that the apostle married was Khadīja....As a virgin she had married 'Utayyiq ibn 'Āyid ('Ā'idh?) (I.H. 'Ābid) ibn 'Abdullāh...and bare him a daughter. Then he died and she married Abū Hāla al-Nabbāsh ibn Zurāra one of Banū 'Amr ibn Tamīm, an ally of Banū 'Abdu'l-Dār and bare him a son and daughter. Then he died and the apostle married her and she bare him his four daughters Zaynab, Ruqayya, Umm Kulthūm, and Fāṭima. After the daughters she gave birth to al-Qāsim, al-Ṭāhir, and al-Ṭayyib. All the boys died while they were at the breast. (In I.K. this tradition comes from al-Zuhrī.)

Ruqayya lived to marry 'Uthmān to whom she is said to have borne a boy called 'Abdullāh (hence 'Uthmān's *kunya*) who died at the breast. When Ruqayya died the apostle gave him Umm

Kulthūm to wife. (The gist of this is given from other authorities by I.K.)

Zaynab was married to Abūl-'Āṣ ibn al-Rabī' and gave birth to Umāma and 'Alī. The latter died young and Umāma lived until 'Alī married her after the death of Fāṭima. After 'Alī was killed she married al-Mughīra ibn Naufal ibn al-Ḥārith ibn 'Abdu'l-Muṭṭalib and died as his wife. (So I.K. *loc. cit.*)

Fāṭima's Marriage

MS. fos. 78b–79a; I.K. III, 346

The story of 'Alī's marriage to Fāṭima is to be found in most of the books of Tradition. Here 'Alī confesses that he feared to ask the apostle for Fāṭima's hand because he had no bridal gift; but the apostle reminded him that he had a cuirass which he had given him and told him to give her that. Thus, as he said, he obtained the apostle's daughter for an article worth a mere four dirhems.

Yūnus reported that he heard Ibn Isḥāq say that Fāṭima bore to 'Alī al-Ḥasan and al-Ḥusayn and Muḥassin. The latter died as a child. She also bore Umm Kulthūm and Zaynab.

'Umar marries Umm Kulthūm Daughter of 'Alī

MS. fo. 79a–b

This and some of the following traditions have nothing to do with the *Sīra* proper and belong in time to the caliphate. After reporting that Zayd and an unnamed girl were born of the marriage, Ibn Isḥāq reports a tradition from 'Āṣim ibn 'Umar ibn Qatāda to the effect that when 'Umar asked 'Alī for Umm Kulthūm's hand 'Alī made the excuse that she was too young. 'Umar replied that that was not the real reason for his hesitation: he wanted to refuse his consent. Let him send her to him that he might see for himself. So 'Alī gave her a robe and told her to go to the commander of the faithful and say "My father says 'What do you think of this dress?'" She went and said these words and 'Umar took hold of her shift. She tried to get it from him saying "Let go!", and he let go exclaiming "A virtuous woman! Go and say to him 'It is fine and beautiful, not at all as you said it was'." So 'Alī married him to her.

A second tradition on the authority of 'Alī ibn al-Ḥusayn says that when 'Umar married Umm Kulthūm he came to an assembly in the apostle's mosque between his grave and the pulpit where only the muhājirs sat; and when they greeted him he told them that the reason for his marrying Umm Kulthūm was that he had heard the apostle say "On the Resurrection Day every tie and relationship will be severed except those that are with me".

There is more than a suspicion of 'Alid propaganda in these two traditions.

Umm Kulthūm marries 'Aun ibn Ja'far ibn Abū Ṭālib

MS. fos. 79b–80a

A tradition which Ibn Isḥāq claims to have had from his father from Ḥasan ibn Ḥusayn from 'Alī says that when Umm Kulthūm was widowed her two brothers Ḥaṣan and Ḥuṣayn came and reminded her that she was the first Muslim lady, the daughter of their greatest lady; and they said that if she gave 'Alī the disposal of her person he would marry her off to one of his orphans. If she wanted to acquire great wealth she could do so. At this point 'Alī entered leaning on his stick. After reminding them, the sons of Fāṭima, of their exalted position as descendants of the prophet and of their superiority in that respect to the rest of his children he asked Umm Kulthūm to place her future in his hands. She replied that she had the same ambitions as other women and would prefer to manage her own affairs. 'Alī said that this was not really her own thought, but the suggestion of these two men and he would never speak to them again unless she did as he asked. They took hold of his garment and begged him to sit down for they could not bear to be estranged from him. They asked Umm Kulthūm to do as he asked and she at once agreed. Thereupon 'Alī said that he would marry her to 'Aun ibn Ja'far who was a young man. He went home and sent her 4000 dirhems and sent to his nephew to consummate the marriage. Ḥasan said that never in his life had he seen such passionate love as she felt for 'Aun. However, he died quite soon and 'Alī again asked her to let him arrange her affairs and he married her to Muḥammad ibn Ja'far. He sent her 4000 dirhems and then allowed Ja'far to consummate the marriage.

Clearly this story is about 'Alī: it is not from him. Ibn Isḥāq then adds in his own name that Muḥammad ibn Ja'far left Umm Kulthūm a widow without having a child by her.

The Marriage of Zaynab Daughter of 'Alī and Fāṭima

MS. fo. 80a

Zaynab married 'Abdullāh ibn Ja'far ibn Abū Ṭālib and bore 'Alī and Umm Abīhā.[1] The latter married 'Abdu'l-Malik ibn Marwān who divorced her and she afterwards married 'Alī ibn 'Abdullāh ibn 'Abbās.

'Uthmān's Marriage

MS. fos. 80b–81a

The apostle was exceedingly jealous on his daughters' account and would not let them marry men who already had a wife who would be a rival.

On the authority of al-Ḥasan it is said that the apostle counselled 'Uthmān's wife to go to all lengths in obedience to her husband's wishes and to live at peace with him.

The Prophet marries Sauda Daughter of Zama'a

MS. fo. 81a; I.H. p. 1001; L. p. 792; I.K. III, 133; V, 294

The MS. states that Khadīja died three years before the hijra. Sauda was the widow of her cousin Sakrān ibn 'Amr, brother of Suhayl, who had married her when she was a virgin. They migrated to Abyssinia and after their return to Mecca Sakrān died as a Muslim. She had no issue by the prophet.

A discussion as to whether Sauda or 'Ā'isha was the first to succeed Khadīja will be found in I.K., who takes the view that the marriage contract with 'Ā'isha preceded that with Sauda though her marriage was not consummated until after the hijra.

He marries 'Ā'isha Daughter of Abū Bakr

MS. fo. 81a–b; I.H. p. 1001; L. p. 792; I.K. III, 133; V, 294

Here Ibn Isḥāq states roundly that the prophet married 'Ā'isha *after* Sauda. This is one of the passages which I.H. rewrote (*v. supra*).

There follows a lively anecdote from 'Ā'isha of her journey to Medina with the emigrants. Her camel bolted with her in a steep

[1] A strange nickname probably bestowed on the woman because she closely resembled her paternal grandmother. She is mentioned by T. II, 1174, and the same nickname is to be found in T. III, 758 and 1357.

pass. She remembered her mother's cry of horror and a man calling to her to let go the halter and how when she did so the camel began to go round and round, as though someone sitting below it was holding it.

He marries Ḥafṣa Daughter of 'Umar

MS. fo. 81b; I.H. p. 1002; L. p. 793

The MS. records that this marriage followed that with 'Ā'isha.

He marries Zaynab Daughter of Khuzayma

MS. fo. 81b; I.H. p. 1004; L. p. 794; I.K. v, 295

The MS. says that she came after Ḥafṣa and that her first husband was either al-Ḥuṣayn ibn al-Ḥārith ibn al-Muṭṭalib or his brother Ṭufayl. She died in Medina the first of the prophet's wives to die. I.K. reports the same tradition. I.H. says that her first husband was Jahm ibn 'Amr ibn al-Ḥārith, her cousin; and her second 'Ubayda ibn al-Ḥārith ibn al-Muṭṭalib. I.K. gives yet another name: 'Abdullāh ibn Jaḥsh ibn Ri'āb.

He marries Umm Ḥabība

MS. fo. 82a; I.H. pp. 144 and 1002; L. pp. 99 and 793

Here the information given in the two contexts in I.H. is assembled somewhat differently.

He marries Umm Salama

MS. fo. 82a–b; I.H. p. 1002; L. p. 793; T. p. 1771

All that is added to I.H. is that Umm Salama's first husband died in Medina of wounds received at Uḥud. The name of one of her children by him is given as Durra (so T.). I.H. says Ruqayya.

There follows a tradition which Ibn Isḥāq got from his father that Sa'd ibn 'Ubāda used to go round with a bowl of food to the house where the apostle was; and whenever the latter proposed marriage to a woman he would add "And the bowl of Sa'd ibn 'Ubāda will come to you every morning".

A composite *isnād* is given for the statement that Umm Salama's son, Abū Salama, married her to the apostle who in turn married him to the daughter of Ḥamza while they were children and they died before their marriage was consummated. The apostle said "Have I made Salama sorry that he married me to his mother?"

Another tradition is that the apostle married Salama in Shawwāl and consummated the marriage in the same month. She said "Spend a week with me". He said "If you wish I will, and then spend a week with your co-wives; or, if you like, three nights and then I will go the rounds to make up for your time". She said "No, three nights".

He marries Zaynab Daughter of Jaḥsh

MS. fo. 82b; I.H. p. 1002; L. p. 793; I.K. v, 300

In place of I.H.'s "Her brother Abū Aḥmad ibn Jaḥsh married him to her...and concerning her God sent down 'So when Zayd had attained what he wished of her We married him to you'"[1] our MS. simply has: "God married him to her"; it adds that she bore him no children and her *kunya* was Ummu'l-Ḥakam. I.K. in his summary quotation from Yūnus omits these words.

He marries Juwayriya Daughter of al-Ḥārith

MS. fo. 82a–b; I.H. pp. 729 and 1002; L. pp. 493 and 793

The MS. brings together two passages that are widely separated in I.H., and adds that Juwayriya was formerly married to her cousin who was known as Ibn Dhūl-Shafr.

He marries Ṣafīya Daughter of Ḥuyayy

MS. fo. 83a–b; I.H. pp. 763 and 1003; L. pp. 514–15 and 793

Here there are merely verbal differences.

He marries Maymūna the Hilālite, Daughter of al-Ḥārith

MS. fo. 83b; I.H. pp. 790 and 1003; L. pp. 531 and 793–4

In I.H. there is a tradition from Ibn 'Abbās that the apostle married Maymūna while a pilgrim under taboo. The MS. has no tradition from Ibn Isḥāq to that effect; instead we have a flat contradiction of what al-Bakkā'ī reported. It runs thus: "A trustworthy person told me that Sa'īd ibn al-Musayyib said 'This fellow 'Abdullāh ibn 'Abbās alleges that the apostle married Maymūna while he was a pilgrim. He lied. The apostle came to

[1] Sūra 33, 37.

Mecca as an ordinary visitor (*muḥill*), and normal conditions (*ḥill*) and marriage go together. Thus he introduced doubts into men's minds.'"

This was a vexed question among Muslim lawyers. There is sharp disagreement voiced by other authorities quoted by Yūnus in his collection which goes to show that there was no agreement in his day.

He marries Asmā' Daughter of Ka'b, and 'Amra Daughter of Yazīd

MS. fos. 83b–84a; I.H. p. 1004; L. p. 794; I.K. v, 300

The MS. says that Muhammad married Asmā', a Jūnī woman, and divorced her forthwith. This seems to be a variant of I.H. where it is said that he married a Kindī woman, Asmā', daughter of al-Nu'mān, and finding that she was leprous sent her away with a present. From another authority Yūnus refers this story to a woman of Banū Ghifār.

The MS. adds that 'Amra belonged to Banū'l-Waḥīd, a clan of Kilāb, and had been married to al-Faḍl ibn al-'Abbās. It omits the story of her reluctance to marry the apostle.

In his summary Ibn Kathīr approves of the chronological order of the marriages given by Yūnus on Ibn Isḥāq's authority.

Umm Ḥabīb Daughter of Ibn 'Abbās

MS. fo. 84a

Here there is a tradition from Ibn 'Abbās that the apostle saw this child crawling about and said that if he were alive when she grew up he would marry her. He died while she was still a child and she married al-Aswad ibn 'Abdu'l-Asad, brother of Abū Salama, and she bore him Rizq and Lubāba. The latter was named after her grandmother, Ummu'l-Faḍl whose name was Lubāba.

The Apostle's Concubines

MS. fos. 84b–85a; cf. I.H. p. 693; L. p. 466; I.K. v, 304

Nine wives survived the apostle. Khadīja and Zaynab, "mother of the poor", predeceased him. Those who had been among the emigrants to Abyssinia were Umm Salama, Umm Ḥabība, and one other.[1] Only Khadīja bore him children. His concubines were Rayḥāna, daughter of 'Amr, and Māria the Copt

[1] *fulāna.* Presumably Sauda is intended, cf. I.H. p. 214; L. p. 148.

who gave birth to Ibrāhīm. She and Khadīja were alone in bearing him children. It is said that Ibrāhīm died at the age of eighteen months and no prayers were said at his burial.

There follows a tradition given in 'Alī's words. (Ibn Kathīr has it in a slightly different form, and it is alluded to by other traditionists.) It records how evil gossip gathered round the repeated visits that Māria's cousin (Ibn Kathīr gives his name as Mābūr) paid to her. The prophet told 'Alī to take his sword and, if he found the man with Māria, to kill him. 'Alī said: "I will be about your business like a hot nail; nothing shall deflect me till I have carried out your command. But will one present see what the absent does not?" The apostle answered "Yes, certainly". 'Alī reported how he took his sword and went and found the man with Māria. He, seeing the drawn sword, knew that 'Alī was making for him and ran to a palm-tree and climbed it. When he was halfway up 'Alī was getting near him, so he threw himself down and fell on his back. As his leg rose in the air 'Alī saw that he was devoid of male organs, so he sheathed his sword and went and told the apostle who gave thanks to God.

How the Apostle was Compensated for Childlessness

MS. fo. 85a–b; I.H. p. 261; L. p. 180

In I.H. there is no *isnād*. Here the tradition is reported by Yazīd ibn Rūmān. I.H.'s note in the second tradition shows that the genealogy of the narrator was defective in al-Bakkā'ī's *riwaya*. The MS. shows that the fault was not due to Ibn Isḥāq.

The Story of the Mockers and the Verses (about them)

MS. fo. 85b; I.H. pp. 271–2; L. p. 187; I.K. III, 195–6

The story here is considerably shorter. There is no mention of the prophet's having cursed al-Aswad ibn al-Muṭṭalib, and the five men are named summarily and dealt with in a different order, but in the same way. The quotation from Sūra 15, 94 comes fittingly at the end.

Hishām ibn al-Walīd Protects his Brother

MS. fos. 85b–86a; I.H. p. 207; L. p. 145

There is no important difference here.

The Story of Rukāna ibn 'Abdu Yazīd

MS. fo. 86b; I.H. p. 258; L. pp. 178–9; I.K. III, 103

Here the story is much shorter and it is to be noted that there is no mention of the prophet's making a tree come to him and go back to its place. Rukāna attributes his defeat in wrestling to Muhammad's sorcery, but that is all.

The Signs of Prophethood

MS. fo. 88a–b

Ibn Isḥāq reports a tradition from Abū Hurayra that the apostle said that he had been told about[1] a man who rode a cow and beat it to quicken its pace. The cow said "My good man, I was not created for this!" The people were astonished and the apostle said "Do you marvel at this? I believe it, and so do Abū Bakr and 'Umar who were not there." He continued "A wolf attacked a man's flock and carried off a sheep, and the man went after it and took it from him. The wolf said 'You've kept it from me today but who will protect it on the day of resurrection[2] when there will be no shepherd but myself?'" The people were astonished, etc., as above.

Why these were signs of Muhammad's prophethood becomes clear only from the reading of I.K. VI, 143 f., where similar stories are told with the added information that the animals testified to his apostleship.

Abū Hurayra of Daus becomes a Muslim

MS. fo. 89b

Ibn Isḥāq on the authority of "one of my companions" reports that Abū Hurayra said "In the pagan period my name was 'Abdu Shams ibn Ṣakhr. In Islam I was named 'Abdu'l-Raḥmān. I got the nickname of Abū Hurayra thus: I used to shepherd a flock of his (the name of the owner is not given) and I found some young wild cats and put them in my sleeve. When I brought his flock back at night he heard their mewing in my bag and asked what they were. When I told him he said 'You are Abū

[1] *'an* could mean "on the authority of".

[2] *yaum al-sab'i.* This is explained in Ibn al-Athīr's *al-Nihāya*, II, 144, and in Lane's *Arabic–English Lexicon*, p. 1296c.

Hurayra' (the father of the kitten) and the name stuck to me." Ibn Isḥāq added that he was a man held in high regard among Daus, hence he must be accounted to belong to them.

'Abdullāh the Mazanī Dhū'l-Bijādayn

MS. fo. 92b; cf. I.H. pp. 904–5

This man lived under the protection of his uncle who used to treat him generously until he heard that he followed the religion of Muhammad. He told him that if he continued to do so he would take back everything that he had given him. He refused to abandon his religion and so his uncle stripped him of everything that he had given him down to his very clothes. 'Abdullāh went to his mother who tore her garment into two pieces: half he used to cover the lower part and half the upper part of his body. Then he went and prayed the morning prayer with the apostle. As the apostle prayed he stared at the people, looking at those who came to him. Noticing this, Muhammad asked him who he was and he said that he was 'Abdu'l-'Uzzā. The apostle replied, "No, you are 'Abdullāh with-the-two-garments. Stay by my house." So he used to remain by the apostle's door. He used to lift up his voice in the recitation of the Koran and the cry of Allāhu Akbar and in giving praise. 'Umar said, "He is...". The apostle answered "Let him alone, for he is one of the...".[1]

In the *Sīra* Ibn Isḥāq relates how the prophet himself helped to bury this man and made a special prayer for him; and Ibn Hishām gives a garbled version of the story of the two garments.

The Night Journey

MS. fo. 92b; I.H. pp. 263f.; I.K. III, 108f.

Here many of the traditions recorded in the *Sīra* are missing. There is the introductory warning that the narrative is a great test of men's faith; the tradition of 'Ā'isha that the prophet's body never left its place, only his spirit being taken by night; Muhammad's description of Jesus and the prophets to his companions; the vessels containing water, wine, and milk which were offered to him to drink; his having seen heaven and hell and various sights in heaven; and his having been given instructions about ritual prayer. The photostat is not clear in places, especially where there are no diacritical marks.

[1] Two words, one of which is written in the margin, are obscure.

Part v ends with traditions about the prophet's failure to respond to greetings while he was at prayer.[1] For the most part they are illegible.

APPENDICES

A

نا احمد نا يونس عن ابن اسحق قال كنت جالسا مع ابى جعفر محمد بن على فمرّ بنا عبد الرحمن الأعرج مولى ربيعة بن الحرث بن عبد المطلب فدعاه فجاءه فقال يأعرج ما هذا الذى تحدث به ان عبد المطلب هو الذى وضع حجر الركن في موضعه؟ فقال اصلحك الله حدثني من سمع عمر بن عبد العزيز يحدث انه حدث عن حسّان ابن ثابت يقول حضرت بنيان الكعبة فكأنى أنظر الى عبد المطلب جالسا على السور شيخ كبير قد عصب له حاجباه حتى رفع اليه الركن فكان هو الذى وضعه بيده. فقال انفر راشدا. ثم اقبل علىّ ابو جعفر فقال ان هذا لشىء ما سمعنا به قط وما وضعه الا رسول الله صلعم بيده

B

نا أحمد نا يونس عن ابن اسحق قال فحدثت ان رسول الله صلعم قال وهو يحدث عن زيد بن عمرو بن نفيل أن كان لأوّل من عاب علىّ الأوثان ونهانى عنها. أقبلت من الطائف ومعى زيد بن حارثة حتى مررت بزيد بن عمرو وهو بأعلى مكة وكانت قريش قد شهرته بفراق دينها حتى خرج من بين أظهرهم وكان بأعلى مكة. فجلست اليه ومعى سفرة لى فيها لحم يحملها زيد بن حارثة من ذبائحنا على اصنامنا. فقرّبتها له وأنا غلام شاب فقلت كل من هذا الطعام أى عمّ. قال فلعلها أى ابن أخى من ذبائحكم هذه التى تذبحون لأوثانكم. فقلت نعم. فقال اما انك يابن أخى لو سألت بنات عبد المطلب أخبرنك انى لا آكل هذه الذبائح فلا حاجة لى بها. ثم عاب علىّ الأوثان ومن يعبدها ويذبح لها. وقال انما هى باطل لا تضرّ ولا تنفع أو كما قال. قال رسول صلعم فما تمسحت بوثن منها بعد ذلك على معرفة بها ولا ذبحت لها حتى أكرمنى الله عز وجل برسالته

C

حدثنا أحمد نا يونس بن بكير عن محمد بن اسحق قال حدثنى عبد الله بن ابى بكر عن ابى جعفر قال كان رسول الله صلعم يصيبه العين بمكة فتسرع اليه قبل ان ينزل عليه الوحى. فكانت خديجة ابنة خويلد تبعث الى عجوز بمكة ترقيه فلما نزل عليه القران فأصابه من العين نحو مما يصيب قالت له خديجة يرسول الله ألا أبعث الى تلك العجوز ترقيك؟ قال امّا الآن فلا.

[1] These traditions agree in substance with Bukhārī, xxi, bāb 15, but they come from other traditionists.

D

نا أحمد نا يونس عن ابن اسحق قال حدثنى يزيد بن زياد عن محمد بن كعب القرظى قال حدثنى من سمع على بن ابى طالب رضه يقول خرجت فى يوم شات من بيت رسول الله صم ولقد أخذت اهابا معطونا فحويت وسطه فأدخلته فى عنقى وشددت وسطى وحزمته بخوص النخل وانى لشديد الجوع فلو كان فى بيت رسول الله صم طعام أطعمت منه فخرجت التمس شيئا. فمررت بيهودىّ فى مال له وهو يستقى ببكرة له فاطلعت عليه من ثلمة فى الحائط فقال مالك يا عربىّ هل لك فى كل دلو بتمرة؟ فقلت نعم فافتح الباب حتى أدخل. ففتح فدخلت فأعطانى دلوه فلما نزعت دلوا أعطانى تمرة حتى اذا امتلئت كفى ارسلت الدلو وقلت حسبى فأكلتها ثم شرعت فى الماء فشربت ثم جئت المسجد فوجدت رسول الله صم.

CPSIA information can be obtained
at www.ICGtesting.com
Printed in the USA
BVHW050943060223
657963BV00009B/661

9 781014 903037